THE

Sacred Cosmos

Christian Faith and the Challenge of Naturalism

CHRISTIAN PRACTICE OF EVERYDAY LIFE

Terence L. Nichols

Brazos Press

A Division of Baker Book House Co
Grand Rapids, Michigan 49516

© 2003 by Terence L. Nichols

Published by Brazos Press
a division of Baker Book House Company
P.O. Box 6287, Grand Rapids, MI 49516-6287
www.brazospress.com

Printed in the United States of America

Library of Congress Cataloging-in-Publication Data
Nichols, Terence L., 1941-
    The sacred cosmos : Christian faith and the challenge of naturalism / Terence Nichols.
        p.   cm.—(Christian practice of everyday life)
    Includes bibliographical references (p   ).
    ISBN 1-58743-046-0 (pbk.)
    1. Apologetics. 2. Naturalism—Religious aspects—Christianity. I. Title.
II. Series.
BT1200.N53 2003
231.7—dc22   2003015445

All Scripture is taken from the New Revised Standard Version of the Bible, copyright 1989 by the Division of Christian Education of the National Council of the Churches of Christ in the USA. Used by permission.

Chapter 5 is a revision of "Evolution: Journey or Random Walls" in *Zygon* 31 (1): 193–210.

Chapter 8 is a revision of "Miracles in Science and Theology" in *Zygon* 37 (3): 703–715.

For Mabel

"A good wife who can find?
She is far more precious than jewels."
—Proverbs 31:10

THE CHRISTIAN PRACTICE OF EVERYDAY LIFE

David S. Cunningham
and William T. Cavanaugh, series editors

This series seeks to present specifically Christian perspectives on some of the most prevalent contemporary practices of everyday life. It is intended for a broad audience—including clergy, interested laypeople, and students. The books in this series are motivated by the conviction that, in the contemporary context, Christians must actively demonstrate that their allegiance to the God of Jesus Christ always takes priority over secular structures that compete for our loyalty—including the state, the market, race, class, gender, and other functional idolatries. The books in this series will examine these competing allegiances as they play themselves out in particular day-to-day practices, and will provide concrete descriptions of how the Christian faith might play a more formative role in our everyday lives.

The Christian Practice of Everyday Life series is an initiative of The Ekklesia Project, an ecumenical gathering of pastors, theologians, and lay leaders committed to helping the church recall its status as the distinctive, real-world community dedicated to the priorities and practices of Jesus Christ and to the inbreaking Kingdom of God. (For more information on The Ekklesia Project, see www.ekklesiaproject.org.)

Contents

Introduction
The Dying of God

In a recent article entitled "God's Funeral," Philip Yancey describes the decline of religion in the Netherlands.

> Dutch Christians told me that a century ago, 98 percent of Dutch people attended church regularly; within two generations the percentage fell into the low teens. Today, it's under 10 percent. Almost half the church buildings in Holland have been destroyed or converted into restaurants, art galleries, or condominiums. . . . For a majority of Europeans, the church seems wholly irrelevant.[1]

For the first time in history, Yancey notes, most people in Europe do not pray or worship. They have lost the sense of a transcendent order, an order of truth, values, and reality beyond themselves. Instead, they must create their own. This, indeed, seems to be the message of post-modernism—all persons have their own version of truth, value, and reality, and no authority can tell them differently.

What Europe has experienced is secularization. Objectively, this means the removal from society of religious influence and symbols. Subjectively, it means what sociologist Peter Berger calls the "secularization of consciousness."[2] Secularized people no longer think much about God, or interpret their lives or the world in religious terms; religious creeds and interpretations have lost plausibility. The most visible indicator of this is the media, which in both Europe and America is almost entirely dominated by secular subjects and analysis: news, sports, food, society, fashion, business, advertisements, and so on.

The paradox of America is often noted. By external indicators, the United States is a very religious country. Americans go to church, pray, and believe in God in large numbers, far greater than Europeans.[3] But American culture seems to be extremely materialistic, as evidenced in the media. How can we explain this? Berger thinks that in America the churches themselves have become largely secularized. Supernatural realities such as miracles, angels, afterlife, a sacred cosmos, and so on are rarely broached, at least in mainline Protestant denominations (and less and less so in Roman Catholic churches). God has become distant from everyday life. People may still believe in God, go to church, even pray, but without deep conviction. Admittedly, this is probably more true of so-called "liberal" denominations than for more conservative evangelical and Pentecostal churches. But it is still a widespread phenomenon, and not just in Christianity. I have a friend, a biologist and an observant Jew, who will not come to the university on Jewish holy days. Accordingly, he must plan his academic calendar carefully. He is deeply committed to Judaism, and to raising his son as an observant Jew. But he does not believe in God (though he prays). He is an agnostic. Furthermore, he claims most of his Jewish friends (presumably in his conservative synagogue) also are agnostics. This is an extreme example, but one finds the same sort of thing (though not so extreme) in college classrooms. Students will often say that they are Catholic or Lutheran, but most know little about their faith, which does not seem to greatly influence their everyday lives. It does not, for example, weigh heavily in their career decisions. Students typically ask: "What do I want to do with my life?" not: "What does God want me to do with my life?" Though I teach at a Roman Catholic university, few of the students enter because it is Catholic. Mostly they come for career opportunities, especially business. They are quite uncertain about supernatural realities, such as an afterlife, and do not have the assurance that characterizes those of deep faith. (There is, however, a countermovement among students who are actively embracing traditional faith, but it is a minority.)[4]

Now it is not news that Western society is becoming secularized. Indeed, the conviction of this book series (The Christian Practice of Everyday Life) is that "Christians must actively demonstrate that their allegiance to the God of Jesus Christ always takes priority over secular structures that compete for our loyalty—including the state, the market, race, class, gender, and other functional idolatries." What is in dispute is the origin of secularization. Berger's book *The Sacred Canopy,* a classic analysis of secularization, concludes that the cause and carrier of secularization is the modern capitalist economic system.[5]

I disagree with this analysis. Certainly the love of money and consumerism are strong vectors of secularism. Scripture itself tells us that "the love of money is a root of all kinds of evil" (1 Tim. 6:10). But greed and the love of money were also problems in medieval and Renaissance times, when society was more religious. The problem is deeper and originates further back—with the separation of God from nature, a split that began in the late medieval and early modern period. This resulted in the (perceived) separation of God from everyday life that is so characteristic of contemporary secular societies. The main carrier of this has been modern natural science. Science came to understand nature as a mechanical system that operated more or less independently from God. God was thus gradually (over centuries) removed from the cosmos. In the process, nature came to be seen as resources to be manipulated and exploited for gain, thus resulting in the industrial revolution and modern capitalism. But industrialism and capitalism were themselves the result of an earlier revolution, the revolution in understanding of God and nature. Ancient and medieval Christians lived in a sacred cosmos and saw nature as a window or sacrament that expressed the beauty, majesty, and glory of God. One finds this in the Psalms, in early Christianity, especially Celtic Christianity, in the writings of St. Augustine, in medieval writers such as St. Bonaventure, down to Romantic nature painting of the early nineteenth century (see chapter 2). Sacraments make God present and invite the believer into a sharing of God's presence. But for a sacrament to work, there has to be some similarity, some unity, between God and the sacrament. This is strikingly expressed in the Orthodox, Catholic, Anglican, and Lutheran conviction that the Eucharist is really the body and blood of Christ (the doctrine of "real presence"). By partaking of the Eucharist, one becomes sacramentally united with God and with Jesus. (This kind of union can of course happen in other ways: in prayer, in reading Scripture, in meditation and contemplation, even in active works of charity.) If nature is seen sacramentally, rather than as an object to be investigated and used, it also can mediate the presence of God. Seen sacramentally, nature is a sacred cosmos, for whatever mediates God's presence is sacred.

What replaced this sacramental sense of nature is the sense that nature is a self-sufficient system governed by its own laws, and that God (if God exists) is separate from nature and the tasks of everyday life. The Catholic Benedictine monk Bede Griffiths, who for decades lived in India, tells the following anecdote. In India, Hindu children, when asked where God is, point to their hearts. But Catholic children in India, when asked the same question, point to the sky.[6] God is "up there" but not "here with us."

This changed perception of God, which I will discuss in the following chapters, is one of the most radical changes that has affected Western society. For the first time in human history, there are large sections of peoples, even whole countries, for whom God is distant and irrelevant (if indeed God exists at all). This sense pervades the cultures of the developed West, including American culture. It is this that makes the practice of Christianity in everyday life so difficult, especially for young people, who grow up enchanted by the secular media. All the other secular threats—materialism, commercialism, consumerism, the religion of fun—stem from this.

I have said that natural science is the predominant carrier of secularized consciousness, but I do not think there is any unbridgeable chasm between faith and science. Far from it: many early scientists were devout Christians, and even today many scientists are. The real problem is not science; it is the philosophy of naturalism, which has captured the allegiance of many scientists, philosophers, and also much of the public. ("Naturalism," as used here, is the belief that nature is all that exists, and that everything can be explained by natural causes and therefore by science. There is no nonmaterial reality, such as God.) But I will argue that one need not embrace naturalism to do good science. Natural science and theology (which is the dimension of religion concerned with understanding)[7] are complementary. Each needs the other for a full comprehension of reality. The real source of secularism, then, and the real challenge to the practice of Christianity in everyday life, is the pervasive philosophy of naturalism.

There are four prominent themes in this book. The first theme explores the challenge of naturalism and the historical background that led up to it (chapters 1 and 2). The second theme is the thesis that modern secularism stems from a developing chasm or dualism between God and creation, so that God has become remote and unimportant to the modern secular world. As an antidote to this, I argue that nature is part of a hierarchy of being, in which created realities express the beauty, majesty, and design of God. In this sense, nature can be seen as a sacred cosmos, a kind of sacrament that makes God present, rather than as a self-sufficient material system that leaves God out of the picture (chapter 3). The third theme is the thesis that Christian thought is a more adequate account of all the evidence (both natural and human) than is naturalism. In this regard, I will focus on four areas: the origin of the universe (the so-called "Big Bang" theory—chapter 4); the evolution of the universe (chapter 5); human nature (chapters 6 and 7); and miracles (chapter 8). The fourth theme is that science and Christian theology are complementary, not conflicting (chapters 9 and 10). The real conflict is between Christianity and naturalism. Theology needs science to understand the world; science needs theology to avoid an idolatry of nature, and even to retain its own inner coherence.

■ The Challenge of Naturalism

Some years ago the astronomer Carl Sagan hosted a popular television series entitled *Cosmos*. Near the beginning he declared: "The cosmos is all that ever was, is, or shall be." By "cosmos" he meant the natural universe, and by the natural universe, what could be detected and measured by natural science. Apart from the cosmos, or nature, there is nothing: not angels, souls, the blessed in heaven, spirits, or God. The natural world is all that exists. Sagan's way of putting it ("was . . . is . . . shall be") echoed the Christian doxology: "Glory be to the Father, Son, and Holy Spirit, as it was, is, and ever shall be, world without end forever and ever." Indeed, his statement expressed an alternate religion and doxology: what is truly awe-inspiring and worthy of reverence is nature itself. Certainly for Sagan, the idea that there could be a dimension of reality more truly real and enduring than nature itself would be nonsense.

Many thinkers, especially in the university world and in the sciences, agree with Sagan's view. Harvard biologist Edward O. Wilson, one of the founders of sociobiology, contends:

> We have come to the crucial stage in the history of biology when religion itself is subject to the explanations of the natural sciences. As I have tried to show, sociobiology can account for the very origin of mythology by the principle of natural selection acting on the genetically evolving material structure of the human brain. If this interpretation is correct, the final decisive edge enjoyed by scientific naturalism will come from its capacity to explain traditional religion, its chief competitor, as a wholly material phenomenon. Theology is not likely to survive as an independent intellectual discipline.[8]

What does Wilson mean here by "scientific naturalism"? "Naturalism," as we have seen, is the philosophical belief that nature is all that exists, and that therefore everything can be explained by natural causes. Supernatural causes, such as God, angels, and the soul, do not exist. By calling it "scientific naturalism," Wilson seems to assume that naturalism is intrinsic to the scientific outlook: science, he implies, tells us that this is the way the world is: only nature exists. This usually goes along with a belief, known as "materialism," that nature is nothing but matter, and that matter is what is detected by scientific measurements.[9] What cannot be measured scientifically, for example, God or the soul, is not part of nature and so does not exist.

Now this idea is not new; it goes back at least to the nineteenth century, and was expressed poignantly by the atheistic philosopher Bertrand Russell almost a century ago:

> That Man is the product of causes which had no prevision of the end they were achieving; that his origin, his growth, his hopes and fears, his loves and beliefs, are but the outcome of accidental collocations of atoms; that no fire, no heroism, no intensity of thought and feeling, can preserve an individual life beyond the grave; that all the labors of the ages, all devotion, all the inspiration, all the noonday brightness of human genius, are destined to extinction in the vast death of the solar system, and that the whole temple of Man's achievement must inevitably be buried beneath the debris of a universe in ruins—all these things, if not quite beyond dispute, are yet so nearly certain, that no philosophy which rejects them can hope to stand.[10]

What has occurred since Russell's time is that many authors have tied naturalism and materialism to evolution, as if a belief in Darwin's ideas about evolution entails naturalism and materialism and rules out religion as a valid worldview. Philosopher Daniel Dennett, for example, declares: "At least in the eyes of academics, science has won and religion has lost. Darwin's idea has banished the Book of Genesis to the limbo of quaint mythology."[11] And Oxford zoologist Richard Dawkins sums up the view he thinks is dictated by evolutionary theory as follows: "The universe we observe has precisely the properties we should expect if there is, at bottom, no design, no purpose, no evil and no good, nothing but blind, pitiless indifference."[12]

Now, the naturalism that these authors are talking about is not the same as naturalism in art or literature; there the term denotes an artistic movement back to nature. Nor is it the same as the Romantic view of nature, exemplified in poets like William Wordsworth or William Cullen Bryant; writers of the Transcendentalist school, such as Ralph Waldo Emerson; or painters of the American Hudson River School, such as Thomas Cole and Frederick Church. For these people, as for the Romantic movement in general, nature is suffused with a transcendent glory, so that it is in effect a sacred cosmos. This movement represents a sacramental view of nature—like a sacrament, nature makes present the divine context in which it is embedded. This is the view that I will defend in this book. But this biblical and traditional view of nature is very different from twentieth-century philosophical or metaphysical naturalism, which Sagan, Wilson, Dennett, and Dawkins proclaim. Their view denies that there is any transcendent reality beyond nature, and is therefore similar to (but not quite identical to) materialism and atheism.[13]

Metaphysical naturalism (hereafter "naturalism" for short) is probably the most serious challenge facing Western Christianity. (It is not so much a problem in areas like Africa and Latin America.) It is not just a challenge to Christianity but to all religions that affirm a reality that transcends nature. (This includes all major religions.) For if nature is all that exists, there cannot be any reality that is greater than and independent of nature. Nor can there be any hope of an afterlife, nor any means to really transcend our natural condition. The consoling grace of God, which frees us from sin, addictions, selfishness, hopelessness, and lovelessness, is, for naturalists, a fiction. As a philosophical attitude, this view of nature has become widespread in academic circles and the media, and in secularized populations generally. It is deeply entrenched in the scientific community.

I will argue, however, that while natural science must be committed to *methodological* naturalism (i.e., its method is to explain phenomena by natural causes, rather than by supernatural causes), there is no reason science or scientists need be committed to *metaphysical* naturalism. Indeed, many of the greatest scientists of the past—Galileo, Newton, Descartes, Pascal, Robert Boyle (one of the founders of chemistry), Johannes Kepler, James Clerk Maxwell, and Max Planck—were Christians or theists (Newton believed in God but denied the Trinity). And many scientists today are also theists (though they are probably a minority). Throughout the book I will lay out a cumulative argument against metaphysical naturalism and for a traditional version of Christianity: one that is able to affirm a transcendent Creator God, miracles, the resurrection and ascension of Jesus, and the existence of an immortal soul and an afterlife, in a way that is consistent with contemporary science. I will defend a theistic version of evolution and contend that the real problem for Christianity is not evolution itself, but a naturalistic view of evolution. Science and theology can be seen as complementary, not antagonistic, but this is only the case if scientists separate their science from the ideology of naturalism. In the end I will hold that there is a kind of unity (not a dualism) between spirit and nature, and that the ultimate purpose of nature is to express, as a kind of sacrament, the divine context that sustains and orders it. My arguments against naturalism will not be based primarily on design, as is the case with the intelligent design movement, but on a sacramental view of nature.

Naturalism is both old and new. As an old challenge, it goes back to the dawn of theism. For the great temptation that confronted the early Israelites was to worship the gods of their neighbors, the Canaanites: deities that were the personifications of natural forces. One could say that right from the beginning of theism, the alternatives have been clear: Do we worship the invisible transcendent God of Judaism, Christianity, and Islam? Or do we put our trust in nature and our own abilities to control it?

As a new and distinct philosophical movement, naturalism goes back to the early twentieth century in America, when it was popularized by the philosopher John Dewey and others. But in the last forty years or so it has become a dominant force in intellectual circles, where it has been voiced by some prominent scientists who have come to believe, largely because of advances in science, that everything will soon be explainable by natural causes alone. What used to be explained by recourse to religion, they say, will soon be explained by natural science. And when science has explained everything, then it will have shown that only nature exists. Thus, the French biochemist and Nobel laureate Jacques Monod wrote thirty years ago: "The ancient covenant is in pieces; man knows at last that he is alone in the universe's unfeeling immensity, out of which he emerged only by chance. His destiny is nowhere spelled out nor is his duty."[14]

Some readers may wonder: If naturalism is such a big threat, why haven't I heard more about it? This is primarily because the term "naturalism" is a philosophical term, used mainly within the academic setting. Within universities, both here and abroad, naturalism is winning the day, certainly among the faculty, and probably among the students. It is hard to estimate the percentage of faculty members who are religious, but it is a decided minority, especially in the sciences. A recent survey in *Scientific American* revealed that 90 percent of the members of the National Academy of Sciences consider themselves agnostics or atheists (among biologists, the figure was 95 percent).[15] Kenneth Miller, a professor of biology at Brown University, sizes up the situation as follows: "It is a fact that in the scientific world of the late twentieth century, the displacement of God by Darwinian forces is almost complete." "Over years of teaching and research in science, I have come to realize that a presumption of atheism or agnosticism is universal in academic life. . . . [I]t would be difficult to overstate how common this presumption of godlessness is, and the degree to which it affects any serious attempt to investigate the religious implications of ideas."[16] Miller is speaking from his perspective as a biologist, but in my experience the situation is not much better in the humanities. Philip Johnson observes that "The definition of knowledge that dominates academia has changed to such an extent that it makes naturalism true regardless of the facts." He then quotes philosopher John Searle as follows:

> The result of this demystification [of religion] is that we have gone beyond atheism to the point where the issue no longer matters in the way it did to earlier generations. For us, if it should turn out that God exists, that

would be a fact of nature like any other. . . . If the supernatural existed, it too would have to be natural.[17]

Finally, well-known British philosopher of religion John Hick writes: "Naturalism has created the 'consensus reality' of our culture. It has become so ingrained that we no longer see it, but see everything through it."[18]

Naturalism is first of all a challenge to religion. If E. O. Wilson's view (quoted above) becomes the consensus among educated persons, then religion might survive among uneducated people, but it would have no intellectual credibility and carry no public conviction. And if what Miller says is true, then the process described by Wilson is already well underway in the universities. It has already made inroads in religion and theology. Consider the following statement of Rudolf Bultmann, one of the great biblical theologians of the twentieth century: "Modern men take it for granted that the course of nature and history . . . is nowhere interrupted by the intervention of supernatural powers."[19]

Naturalism is not just a challenge to religion and theology, however. It also poses a disturbing challenge to traditional philosophical beliefs, such as belief in moral laws, the freedom of the will, and personal responsibility. Philosopher Will Provine declares:

> Modern science directly implies that there are no moral or ethical laws, no absolute guiding principles for human society. . . . We must conclude that when we die, we die, and that is the end of us. . . . Finally, free will as it is traditionally conceived—the freedom to make uncoerced and unpredictable choices among alternative courses of action—simply does not exist. . . . There is no way that the evolutionary process as currently conceived can produce a being that is truly free to make moral choices.[20]

As one of my students, a biology major, put it to me, morality is like the rules developed by wolf packs, genetic drives whose purpose is to enhance what Richard Dawkins calls "selfish genes." In this perspective, even altruism and compassion are really selfishness; they are genetically driven behaviors whose aim is to enhance the survival of the genes one has in common with others. Morals are genetically driven conventions, nothing more.

The denial of free will is particularly problematic for many naturalists since they are committed to freedom, especially academic freedom. But I think Provine is correct that strict naturalism leaves little room for free will. Marvin Minsky, one of the fathers of artificial intelligence, expresses this conflict poignantly:

According to the modern scientific view, there is simply no room at all for "freedom of the human will." Everything that happens in our universe is either completely determined by what's already happened in the past or else depends, in part, on random chance. Everything that happens in our brains depends on these and only on these: A set of fixed, deterministic laws. A purely random set of accidents . . .

Does this mean we must embrace the modern scientific view and put aside the ancient myth of voluntary choice? No. We *can't* do that: too much of what we think revolves around those old beliefs. . . . No matter that the physical world provides no room for freedom of the will: that concept is essential to our models of the mental realm. Too much of our psychology is based on it for us to ever give it up.[21]

As Minsky fears, with the demise of freedom of the will goes also personal responsibility, which is based on the idea that persons know the difference between right and wrong and have the capacity to choose between them. Most of our legal and political system is based on the notion of personal responsibility. If persons really have no freedom and cannot be held responsible, can they then be trusted with political freedom, as in modern democracies? Logically, no; rather, they should be governed by force, like animals, a point argued long ago by Thomas Hobbes.

This leads to another probable consequence of naturalism. Modern democracies are built upon the idea that persons are free and have inborn human rights. This was memorably expressed in the U.S. Declaration of Independence: "We hold these truths to be self-evident, that all men are created equal, that they are endowed by their creator with certain inalienable rights, that among these are life, liberty, and the pursuit of happiness." As the framers of the declaration realized, *inalienable* rights cannot be conferred by the state or by mere convention; they must have a higher source: God. What kind of a case can we make for human rights, if human beings, like other animals, are the accidental products of a blind materialistic process? I do not see how a widespread belief in inalienable rights can be long maintained in such a perspective, even if, with Minsky, we allow their "psychological utility." After all, ancient societies, which had no conception of God-given rights, rewarded the elite and enslaved the rest. The logical thing to do, in an overpopulated world with scarce resources, is to "cull the herd." Get rid of the genetic misfits, who are a drain on society and resources, for the benefit of the genetically well-endowed. Already we are seeing this happen with widespread abortion, and proposals for legalized infanticide are now being made, for example, by Princeton ethicist Peter Singer. Singer advocates a kind of naturalistic utilitarianism. He proposes that "disabled" infants, i.e., those who are malformed, retarded, or medically unfit, should be given a trial period to see if they are genetically fit. If they are not, they should be killed.[22] This is not yet law, of course. Memories of Hitler's euthanasia programs are still too fresh to allow such proposals to pass into law. But the naturalistic logic is compelling: genetically unfit persons will be a drain on the society and will pollute the gene pool if allowed to breed. Just as herdsmen cull unfit animals from the herd, so should we do with humans. In a naturalistic world where religion is only vestigial, given time, such thinking will probably prevail.

Finally, I think there is a kind of *practical naturalism* that is widespread among the U.S. population, especially the young. This is the attitude that maybe God exists, maybe not, but we can't be sure; what we can be really sure about is that nature exists, that nature is matter (what we can touch and feel), and that it is the scientists who tell us about nature. Science proves things, students say, whereas philosophy and especially theology are just opinions. When the scientists speak, people listen; when the bishops speak, some people listen, but most do not. It is the scientists who have become the intellectual authorities in our culture. This is not because the population in general understands much about science; they don't, but they do realize that science works. It has cured diseases, brought us electricity, atomic energy, television, computers, and so on. For practical people, that is what counts. There is a growing backlash against science (which is actually, I believe, a backlash against naturalism). But judging from years of college teaching, I am convinced that science has largely displaced all other forms of knowledge as "real knowledge," even for people who know little about actual science. And the spillover from this is practical naturalism.

Now, practical naturalists may go to synagogue or church; they may pray, they may consider themselves as more or less religious. They are likely to be ethical. But deep down these people are not sure about any transcendent religious reality, especially an afterlife. As a consequence, such persons live their life as if this life and nature are all we can be sure of. This form of naturalism is difficult to measure, because if asked in a poll "Do you believe in God?" or "Do you pray?" such people may say, "Yes." But their lifestyles belie what they say. I suspect that this is the majority of theists in this country. In America, such people often go to synagogue or church; in Europe they usually don't.

Practical naturalism is more benign than "hard" metaphysical naturalism, but is far more widespread. I am convinced that it lies behind most of the secularism and consumerism that is so prevalent in this country. Indeed, secularism and naturalism are virtually synonyms. If this world is all we can be sure of, we had best enjoy it. Paradoxically, the really committed metaphysical naturalist, say a Marxist or a sociobiologist, might be personally more ascetic than the common person, precisely because he or she is a person whose primary interest is in ideas, not things. But these ideas, spread abroad in the general population, still produce secularizing effects.

It might be argued that this is too extreme and bleak a picture. Many people who would consider themselves naturalists would argue that consciousness and freedom are emergent qualities of the mind, which is nonetheless wholly dependent on the brain, and cannot survive without it. Therefore we need not surrender free will or morals or human rights in a naturalistic world. Now certainly there are persons who are metaphysical naturalists who act morally, support freedom and human rights, and so on. I recall a series of conversations I had with a colleague in biology, in which he argued that we are moral because it makes us feel good to be moral. He shares money with the poor, as do I; he would say we do this because we want to. It makes us feel good. Furthermore, many might argue that culture, not biology, is what really determines human behavior, and that our culture, based as it is on human rights, freedom, and morality, will continue even if the biological basis for these beliefs has eroded.

But this is dubious at best. Most people who think this way have been raised in a theistic culture (my biology colleague is Jewish) that grounded values, freedom, and rights in a theistic paradigm. And there is a cultural lag; people remain moral for a few generations because they were raised that way. But only if it doesn't cost them too much—they will be moral as long as it feels good, and does not cost them their lives, jobs, money, social respectability, and so on. But really moral behavior, like standing up to the Nazis when the penalty was death, will be too costly. Gradually, I think the values and rights in a post-theistic culture will erode until they, too, are merely vestigial. We have seen this trajectory in Nazism and in Soviet Russia. If naturalism carries the day here in America, there is no reason to think that it will not follow here too.

These ideas are not new. Centuries ago they were poignantly expressed in the Wisdom of Solomon (2:1–3; 6–11):

> For they reasoned unsoundly, saying to themselves,
> "Short and sorrowful is our life,
> and there is no remedy when a life comes to its end,
> and no one has been known to return from Hades.
> For we were born by mere chance,
> and hereafter we shall be as though we had never been,
> for the breath of our nostrils is smoke,
> and reason is a spark kindled by the beating of our hearts;
> when it is extinguished, the body will turn to ashes,
> and the spirit will dissolve like empty air.
> . . . .
> Come, therefore, let us enjoy the good things that exist,

18

and make use of the creation to the full, as in youth.
Let us take our fill of costly wine and perfumes,
and let no flower of spring pass us by.
Let us crown ourselves with rosebuds before they wither.
Let none of us fail to share in our revelry;
everywhere let us leave signs of enjoyment,
because this is our portion, and this our lot.
Let us oppress the righteous poor man;
let us not spare the widow
or regard the gray hairs of the aged.
But let our might be our law of right,
for what is weak proves itself to be useless."

■ The Christian Response to Naturalism

The ultimate difference between naturalism and theism is one of context. The theist maintains that nature exists within a wider context, God, to whom nature owes its origin and its continuing existence. The naturalist, by contrast, maintains that there is no such ultimate context; the ultimate context is nature itself.

Seen within a divine context, nature looks very different than it does to the naturalist. The universe did not "just happen," the product of a chance fluctuation, as in naturalism. Rather, in a theistic view, there is a sufficient reason for its origin and design (see chapter 4). The development of complex, living beings, including humanity, rather than being a blind, meaningless process, devoid of purpose, hope, or fulfillment, can be seen as the long process of creation, ending in the fulfillment of the kingdom of God (see chapter 5). The complex and paradoxical nature of humanity is not just the accidental byproduct of a random and meaningless process, as Harvard anthropologist Stephen J. Gould observes:

> The origin of *Homo Sapiens,* as a tiny twig on an improbable branch of a contingent limb of a fortunate tree, lies well below the boundary [of law and contingency]. In Darwin's scheme, we are a detail, not a purpose or embodiment of the whole—"with the details, whether good or bad, left to the workings of what we call chance."[23]

Rather, within a divine context, we can explain puzzling features of human life, such as the widespread human conviction that there are indeed moral absolutes that are wrong to transgress, that human beings have the freedom to decide between alternatives, that there is an ultimate fulfillment through religion in an afterlife, that meaning, purpose, beauty, and righteousness may be more important than life itself.

In the end, the problem with naturalism is that its view of nature and of humanity is too limited. Nature and humanity are seen as nothing but complex forms of lifeless matter, destined to dissipate in the ultimate heat death of the expanding universe, as the stars burn out leaving only black holes and empty space. In the theistic view of nature and humanity, however, human nature has the capacity to be elevated by the grace of God, and so also does physical nature. The ultimate destiny of the human is to participate in the divine life, and to be exalted in the resurrection. This is also the destiny of nature, whose fulfillment is to be a sacramental sign of its divine context. This can be approached, but not consummated, in this life.

By itself, nature is a text that is ambiguous and not fully explicable. Its promise of beauty and fulfillment is not fulfilled in this life. Life gives way to death, and beauty to decay, both for all creatures and for nature as a whole. Nature only really makes sense if understood as existing within a larger context: God, the ultimate context of both nature and human life. Otherwise it is like an enigmatic inscription, in a language only partially intelligible, from a forgotten culture. That nature exists within a larger context was the gist of the old idea of a hierarchy of being, but today the very idea of hierarchy, connoting as it does domination and patriarchy, is so hopelessly distorted that it is better to recast the idea in contemporary terms. Thus I will try to explain it along the lines of more and more inclusive and nested contexts. Much of the book will involve explaining God as an active context, and evolution and human nature within a divine context.

■ Christianity and Science

Natural science is dedicated to explaining events according to natural causes, and natural causes are those that can be quantified and measured, and tested by experiment. Thus the first sentence of the preface to the first edition of Isaac Newton's magnum opus, the *Principia* (*The Mathematical Principles of Natural Philosophy*, 1687), reads:

> Since the ancients . . . esteemed the science of mechanics of greatest importance in the investigation of natural things, and the moderns, rejecting substantial forms and occult qualities, have endeavored to subject the phenomena of nature to the laws of mathematics, I have in this treatise cultivated mathematics as far as it relates to philosophy.[24]

Whatever, then, cannot be measured mathematically or empirically tested falls outside of the scientific method. But this did not mean that early scientists thought that supernatural causes did not exist. At the end of the *Principia* Newton appended a "General Scholium" or gloss, in which he states: "This most beautiful system of the sun, planets, and comets could only proceed from the counsel and dominion of an intelligent and powerful being."[25] Practically all of the founders of modern science would have agreed with him. For them, science explained the workings of nature according to natural principles, but natural science could not explain everything—it was a limited endeavor. Nature existed within a divine context. The difference between them and modern scientific naturalists is that the latter think that science can explain everything. But if there is a spiritual dimension to reality, as all the great religions hold, then the methods of natural science, based as they are on mathematics and experiment, cannot reveal the whole truth about reality. Natural science must be complemented by other ways of investigation and other ways of knowing: philosophy, artistic intuition, ethical knowledge, and theology.

Theology is the discipline that attempts to understand divine reality. Theology and science might seem to be at opposite poles of the spectrum. However, as John Polkinghorne argues persuasively in many works, both natural science and theology aim at true knowledge of reality, yet explore different dimensions of reality. Each has its own proper methods and its own goals. What is essential is that each discipline recognize its own limits, and not advance triumphalistic claims that usurp the authority of other disciplines. Theology did this in the Galileo case, but now such triumphalistic claims are being made by some scientists, for example, Sagan, Wilson, and others who do not think that there is any area of reality that corresponds to theology, and that theology and religion are therefore illusions.

If, however, science limits itself to a methodological naturalism and eschews metaphysical naturalism, then there need be no conflict between science and theology. Indeed, since they both are seeking the truth about reality, but are examining different areas, they are not conflicting but complementary. They are sister, or perhaps, cousin, disciplines. But their methods differ, since the dimensions of reality that they investigate differ. As Aristotle noted centuries ago, one must use a method that is appropriate for the nature of the reality one seeks to investigate. Theological methods do not work for the investigation of nature (as the Galileo episode showed), but neither do scientific methods work for the investigation of properly theological subjects, such as the nature of God, God's relation to us, and the nature of spiritual reality (including the spiritual dimension of human beings).

It is an old canard that religion has always opposed scientific progress, and that the triumph of science will also bring the triumph of naturalism and the defeat of religion. This is what E. O. Wilson says, but one can find similar sentiments expressed by many authors. I will argue that just the reverse is the case. Modern natural science only emerged within a theistic civilization, and there is good reason for this. The idea that there were objective laws of nature, which were rational and so could be understood by reason, was grounded in the belief that the creator of nature was an intelligent and rational lawgiver. Once the belief in a rational lawgiver who sustains the whole universe is dissolved, the belief in the universality of objective, rationally knowable laws may well dissolve also, and be replaced by a cacophony of fragmented worldviews characteristic of polytheism, in which one person's perspective is just as valid as another's. We see this already emerging in deconstructionism and postmodernism. Further, if human beings are nothing but the products of their genes and blind deterministic forces, then their ideas also are determined by these forces, and so there is no reason that anyone should accept another's ideas as true. Francis Crick, one of the discoverers of DNA, wrote, "The Astonishing Hypothesis is that 'You', your joys and your sorrows, your memories and your ambitions, your sense of personal identity and free will, are in fact no more than the behavior of a vast assembly of nerve cells and their associated molecules."[26] But if this is true, then all of our ideas, including Crick's, are also determined by the chemical interactions going on in our brains. And if this is so, why should anyone's idea be true, that is, correspond to reality? Once this realization becomes widespread, science, which is above all a system of knowledge, which lives or dies by the truth of its claims, is finished.

Again, the practice of modern natural science depends entirely on a stable society, in which certain foundational values are present. Absent these values, particularly honesty, the whole scientific enterprise collapses. But naturalism tends to undermine these (and all) values. The only real value in naturalistic evolutionary theory is sheer survival, which itself justifies any other value, including falsifying scientific reports so as to advance one's career.

In the end, we need both a scientific vision of reality and a theological vision of reality to retain a balanced viewpoint and a balanced civilization. I do not think these visions will ever be fully united; rather, they will remain complementary. Both an objectivist view of nature and a sacramental view are possible and authentic, and both are true. Neither need exclude the other. Human beings can be seen as both physical systems and as spiritual beings. Both are true, and need not exclude the other. They are not identical, but complementary.

God and Nature
Historical Background

■ This chapter will sketch out the changing views of God and nature from biblical times to the present, as a background for the emergence of naturalism and the modern problem of relating God and nature. Beginning with the Hebrew Scriptures, we find that God was conceived as transcendent over nature (Genesis 1), but also as reflected in and as working through nature (cf. Ps. 19; 147). However, the actions of God were not clearly differentiated from events caused by natural laws: rain, drought, snow, and frost were thought of as due directly to God (Ps. 147: 8, 16). Later, Christians still thought of God as reflected in nature, and as working immanently through nature. However, beginning with the late medieval period, the order of nature, governed by natural causes (so-called "secondary causes"), and the supernatural order, governed by God (the "primary cause"), were sharply distinguished. At the same time, the ancient conviction that nature reflected and expressed God was gradually replaced by the belief that nature was an object controlled by God's will, rather than a kind of sacrament that reflected and expressed God's being. One of the great changes occasioned by the scientific revolution was to conceive of nature as a kind of machine, and God as external to nature. This has led to the prevalent modern view (in the West) of God being marginalized from a wholly natural and secular world.

This story is relevant for this book because I will try to make a case for recovering the old idea of a sacramental unity between God and nature. This unity is not based simply on design, but on a hierarchy of being, in which the fullness of being, God, is reflected and expressed through created beings. But at the same time, we cannot go back to a simple biblical vision of the world, which does not differentiate the actions of God and the actions of nature. We cannot ignore our scientific heritage and the well-developed system of natural causes that it has articulated in recent centuries. What we can do is to try to recover a complementary vision of reality that makes room for both God and nature without collapsing one into the other, thus preserving the proper roles and uniqueness of each.

■ The Hebrew Scriptures

What is unique about the religion of Israel is its affirmation of one God who creates and transcends nature, and who acts in and through nature. This differed radically from the polytheistic religions of Israel's neighbors: the Babylonians, Egyptians, and Greeks, as well as the Canaanites, who were displaced by the invading Israelites after the Exodus. (There may have been a brief flirtation with monotheism under the Egyptian pharaoh Ikhnaton [1377–1358 B.C.], who inculcated the worship of Aton, represented by the sun disc. But this was short-lived; later pharaohs attempted to efface Ikhnaton's reforms and memory from Egyptian records.) The gods of polytheistic religions tended to be personifications of natural powers. This was the ancient version of naturalism, and was the primary challenge to the worship of Yahweh. Ancient Greeks, for example, worshiped Zeus, the mountain god of thunder, storm, and rain; Poseidon, the god of the sea; Dionysius, the god of vegetation; Artemis, associated with trees, springs, rivers, and fertility; as well as a host of river gods, nymphs, and local deities. Canaanite religion worshiped Baal, the god of storm, rain, and fertility, and Astarte, the goddess of fertility (important to farmers in an arid land). Canaanite ritual incorporated sexual rites into the worship of Baal and Astarte. The union of the priest of Baal and his sacred consort, as well as the union of worshipers with temple prostitutes, ensured the continuing fertility of the land. The great temptation of the Israelites, in the early periods of their history, was to combine the worship of Yahweh with the worship of Baal and Astarte. People would sacrifice to Baal and Astarte to bring about fertility of the land, and to Yahweh in times of military crisis. But the message of the Israelite prophets was: Yahweh will not be worshiped alongside other gods. To worship other gods is to fall into idolatry and to abandon Yahweh's covenant. This temptation perdured throughout much of Israel's early history. The destruction of the northern kingdom

in 722 B.C. by the king of Assyria was ascribed by the author of 2 Kings
(17:7–16) to Israel's worshiping gods besides Yahweh:

> This [the fall of Samaria] occurred because the people of Israel had sinned
> against the LORD their God. . . . They rejected all the commandments of the
> LORD their God and made for themselves cast images of two calves; they
> made a sacred pole, worshiped all the host of heaven, and served Baal.

Literary evidence for Israelite nature worship is supported by archaeological evidence, which has shown that in the outlying regions of Israel,
people possessed small statuettes of the goddess of fertility, Astarte.[1] Thus
the temptation to trust in the personified powers of nature rather than
in Yahweh was the greatest challenge to early Yahwism. In this sense,
naturalism was an ancient as well as a modern challenge to theism.

In the Hebrew vision, all of nature is created by God and is under
his complete control. The sun and moon, deified in other cultures, are
called "lights" in Genesis 1:14 and are placed by God in the heavens. To
the Hebrew mind God works directly through the powers of nature:

> He sends out his command to the earth;
>   his word runs swiftly.
> He gives snow like wool;
>   he scatters frost like ashes.
> He hurls down hail like crumbs—
>   who can stand before his cold?
> He sends out his word, and melts them;
>   he makes his wind blow, and the waters flow. (Ps. 147:15–18)

There is in Israel virtually no notion of nature having its own laws apart
from Yahweh. In fact, there is no word for "nature" at all in Hebrew or
cognate languages. What holds nature in being and checks its dissolution
is the controlling will and wisdom of Yahweh. John McKenzie writes:

> The Old Testament conception of nature is that it exhibits the constant
> activity of Yahweh directed by His providence towards men. The whole
> of nature is personalized, and there is no idea of natural causes acting
> according to constant laws and fixed principles. The course of nature
> exhibits the unpredictability which we attribute to personal beings.[2]

In turn, nature, in its order and majesty, testifies to God's majesty, wisdom, and power:

> The heavens are telling the glory of God;
>   and the firmament proclaims his handiwork. (Ps. 19:1)

> O LORD, how manifold are your works!
>> In wisdom you have made them all;
>> the earth is full of your creatures.
>
> . . . .
>> . . . when you take their breath away, they die
>> and return to their dust.
> When you send forth your spirit, they are created;
> and you renew the face of the ground. (Ps. 104:24, 29–30)

The view of God in Hebrew poetry (Psalms, Job, etc.) can be characterized as an early version of what we would call now a sacramental view of nature, in that nature is seen as expressing many aspects of God. This goes beyond simply manifesting God's design. Nature also expresses God's goodness (Genesis 1), God's majesty and glory (Psalm 8; 19), God's permanence (Psalm 18; God as rock, Psalm 90), God's power (Psalm 18; 29), and God's presence (Psalms 8, 29, 140). At the same time, a sacramental view presupposes that God transcends nature. It differs therefore from pantheism, in which the divine is identified with nature, or animism, in which nature itself is seen as animated by spirits or nature deities.

In the Hebrew Scriptures, then, nature is not worshiped. Since it has been created, it is dedivinized. Only Yahweh is divine and worthy of worship. But nature witnesses to the glory and wisdom of God, its creator. Nature is not yet seen as having an independence or lawfulness of its own apart from God. For ancient Israel, human control of nature is indirect: it consists in observing the covenant, in having right relations with God. If these relations are good, nature will be ordered and beneficent; if they are bad, the order of nature, which is held in being by God, will collapse into dissolution and chaos (e.g., the Genesis flood). It is therefore sin and righteousness (not science) that is the key to controlling nature.

The dedivinization of nature proved to be important in the development of natural science. For as long as nature was understood to be under the control of various deities, science, which is based on the idea that nature has fixed laws, could not develop. Nature religions were the principal opponents of monotheistic religion. But they also had no view of lawlike cause and effect, and so would have prevented the development of any physical science.

■ The New Testament

Jesus' teaching in Matthew 6 is similar to that of the Hebrew Scriptures—God works through nature, which expresses God's power:

Therefore I tell you, do not worry about your life. . . . Look at the birds of the air; they neither sow nor reap nor gather into barns, and yet your heavenly Father feeds them. . . . And why do you worry about clothing? Consider the lilies of the field, how they grow; they neither toil nor spin, and yet I tell you, even Solomon in all his glory was not clothed like one of these. (Matt. 6:25–29)

On the other hand nature can become disordered like the waters of chaos. When Jesus stills the storm on the sea (Mark 4:35–41) he rebukes (Greek: *epitimao*) the wind, just as he rebukes demons, and he says to the sea, "Peace! Be still!" just as he commands demons. The miracles of Jesus represent the triumph of the kingdom of God over the forces of disorder. The healing miracles and the nature miracles are, then, a pushing back of the frontiers of the kingdom of Satan.[3]

The New Testament vision is that Christ has reconciled humanity, separated from God since Adam, to God. God's will, "as a plan for the fullness of time, [is] to gather up all things in him [Christ], things in heaven and things on earth" (Eph. 1:10). The culmination of this plan is the resurrection of Jesus, and the subsequent resurrection of all humanity.

The resurrection of Jesus is of course a bodily resurrection—that is the meaning of the empty tomb. It is not the survival of a ghost or a disembodied spirit; the Gospels are at pains to make this clear (cf. Luke 24:39–40; John 20:27). But at the same time the resurrection is not a resuscitation, like the raising of Lazarus. The resurrected Jesus appears in a room where the doors are locked (John 20:26), and disappears before the disciples as they are eating at Emmaus (Luke 24:31). Generally, Jesus' resurrected body does not seem to be constrained by our dimensions of space, though it can appear within them. The body that is raised, according to Paul (1 Cor. 15:44), is not a physical (*psychikon*) body but a spiritual (*pneumatikon*) body, not perishable like earthly bodies.

This is a paradox. Paul, in trying to describe the nature of the resurrected body, argues that there are different kinds of "flesh" (*sarx*) and different kinds of heavenly and earthly bodies (*somata*):

Not all flesh is alike, but there is one flesh for human beings, another for animals, another for birds, and another for fish. There are both heavenly bodies and earthly bodies, but the glory of the heavenly is one thing, and that of the earthly is another. . . . So it is with the resurrection of the dead. What is sown is perishable, what is raised is imperishable. It is sown in dishonor, it is raised in glory. It is sown in weakness, it is raised in power. It is sown a physical body, it is raised a spiritual body. . . . For this perishable body must put on imperishability, and this mortal body must put on immortality. (1 Cor. 15:39–53)

I read Paul here as arguing that even in the natural world there are different kinds of matter. Therefore the possibility of a resurrected "spiritual" body, which differs from ours in the kind of matter it has, makes sense; the same pattern can be seen in nature. But this argument no longer makes sense in our modern context, because all matter is understood to be made up of the same stuff: atoms and molecules. This is probably the biggest reason that the resurrection seems incredible today. Our modern cosmos has no place for anything like "spiritualized" matter. I will argue in a later chapter, however, that there may be possibilities for higher states of matter even in modern cosmology.[4]

Not content with arguing for the resurrection of Jesus, Paul also argues that Jesus' resurrection is a pledge—the first fruits—of the resurrection to come for "those who belong to Christ" (1 Cor. 15:20–28). He goes on to claim, in a famous passage, that the resurrection will extend to all nature as well.

I consider that the sufferings of this present time are not worth comparing with the glory about to be revealed to us. For the creation waits with eager longing for the revealing of the children of God; for the creation was subjected to futility, not of its own will but by the will of the one who subjected it, in hope that the creation itself will be set free from its bondage to decay and will obtain the glorious freedom of the glory of the children of God. (Rom. 8:18–21)

This vision of a creation that will be transformed in the end times (the *eschaton*) accords with the Old Testament hope for a messianic age, and with the vision of a new heaven and a new earth in the book of Revelation: "Then I saw a new heaven and a new earth; for the first heaven and the first earth had passed away" (Rev. 21:1). This belief, like that in Jesus' resurrection, would seem to require a different kind of materiality, and a transformed kind of nature, for its fulfillment.

As in the Hebrew Scriptures, so also in the New Testament, there is no sense of nature having its own autonomy. Nature, according to Paul, is in an imperfect state because of human sin. Even death is not natural but is a result of sin. Conversely, to the wholly righteous one (Jesus), nature is entirely subject—even the winds and the sea obey him. And a restoration of human oneness with God, after the resurrection, will result in a restoration of nature also.

■ The Patristic Period

Early Christian authors displayed two very different attitudes toward nature. One, under the influence of neo-Platonism, was a world-denying and nature-denying tendency, which saw nature and the body as hindrances to the ascent toward God; the other was a tendency to see nature as the manifestation of God's glory. The Alexandrian theologian Origen (185–253) represents the first of these. Celtic Christianity represents the second. Some authors, such as Augustine, display both tendencies.

Origen

The theology of Origen was strongly marked by the philosophy of Neoplatonism. This philosophy viewed the world as a hierarchical system, with the One (God), the source of creation (*hieros arche* = sacred source), wholly unified and spiritual, existing at the top of the hierarchy, and the rest of creation ("the many") existing as descending levels of being of increasing materiality. In the Platonist scheme, the generation of the world occurred by a process of emanation, the overflowing of the superabundance of the goodness of the One. A passage from a fifth-century A.D. author, Macrobius, expresses the Platonist hierarchy well:

> Since, from the supreme God Mind arises, and from Mind, Soul, and since this in turn creates all subsequent things and fills them with life, and since this single radiance illumines all and is reflected in each, as a single face might be reflected in many mirrors placed in a series; and since all things follow in a continuous succession, degenerating in sequence to the very bottom of the series, the attentive observer will discover a connection of parts, from the supreme God down to the last dregs of things, mutually linked together and without a break.[5]

What differentiates Neoplatonism from Christianity (and Judaism) is the doctrine of creation: the world in the latter does not emanate or overflow from God; it is created by an act of God's will. And the world is good because created by God; it reflects God's own goodness. Yet in some ancient Christian authors, such as Origen, the Neoplatonic idea of the inferiority of matter affects their view of the world.

Thus the notion of a hierarchy of being can lead in two directions. One is an emphasis of the beauty, goodness, and variety of creation as an expression of the infinite good of God. We will encounter this in Augustine. The other is to emphasize that what is really real is the spiritual: the One and its manifestations (Mind, identified by Christians with the Logos or Christ, and Soul). The material world, since it constantly changes and passes out of existence, is less real than the spiritual world. The human being, composed of a material body and a spiritual soul, bridges the material and the spiritual worlds. But the body ages, degenerates, and dies; only the spiritual portion of the human person survives. Therefore the interest of serious Christians was not in the world that perishes, but in the spiritual world that does not perish. There is a tendency in this view of the universe to devalue the material world in favor of the spiritual. The predominant metaphor for spiritual perfection becomes that of ascent from lower things to higher, imperishable, and eternal things. In extreme forms, this can result in a contempt for the body (expressed in exaggerated asceticism) and for nature. Both of these are evident in the writings of Origen.

Origen thought that the material world originated as a result of a fall from spirituality to materiality. In the beginning God had created pure intelligences, all equal, with ethereal bodies. All except Christ became cold in their love of God, and fell away, some to become angels, some to become human, some to become demons. The material world was created to provide corporeal bodies for these fallen spirits, as well as the possibility for purgation, through trials and suffering, and so for a return to their spiritual state. Animals Origen saw not as fallen spirits, but as created for the edification of humanity. This also is the purpose of the whole natural world: "The Creator, then, has made everything to serve the rational being and his natural intelligence."[6] In a striking metaphor, Origen compared nature to the afterbirth of the human birth. Origen believed in the resurrected body, but in his conception this body was so spiritualized as to have almost lost its corporeality.

Certain tenets of Origen's theology were later condemned, but its spiritualizing influence continued on in the Christian tradition. He stands as a signal representative of the side of Christianity that tends to see nature (and the body) as a degraded form of being, having little ultimate value of its own.

Celtic Christianity

At the opposite end of the spectrum from Origen and Alexandrian Christianity is Celtic Christianity, which flourished in the British Isles from about 400 to 1000 A.D. It is associated with the names of saints such as Patrick, Columba, Cuthbert, Columbanus, and Brigid, and with monastic Christianity. The centers of Celtic Christianity were not cities (as was the case with Mediterranean Christianity), but monasteries, frequently founded in remote locations, such as the great monastic centers on the islands of Iona and Lindisfarne. From these monasteries missionaries ventured forth, spreading Christianity throughout the British Isles and even onto the European continent. St. Columbanus, one of the most famous of the Irish missionary monks, founded monasteries in France, Switzerland, and Italy, where, in 614 (at the age of 74), he established Bobbio, which became one of Italy's largest monasteries. Celtic Christianity was decentralized, since it was not based in cities, but in tribes and abbeys. After the Synod of Whitby (664) it was very gradually displaced by a more organized, urban, and Roman-centered form of Christianity.

The Celts were originally a tribal people who had a strong sense of the presence of deities and spirits in nature. This sense of the divine in nature carried over into their Christianity. In the words of Saunders Davies, "For the Celt creation is translucent; it lets through glimpses of the glory of God."[7] Much of their legacy exists now in poems and sacred songs, such as the following:

> There is no plant in the ground
> But is full of His virtue
> There is no form in the strand
> But is full of his blessing.
>
> There is no life in the sea
> There is no creature in the river
> There is naught in the firmament
> But proclaims His goodness.[8]

When asked by the daughters of the King of Tara where his God had his dwelling, St. Patrick is said to have replied:

Our God is the God of all men, the god of heaven and earth, of sea and river, of sun, moon, and stars, of the lofty mountains and the lowly valley, the God above heaven, the God in heaven, the God under heaven. He has his dwelling around heaven and earth and sea and all that is in them. He inspires all, he quickens all, he dominates all, he sustains all. He lights the light of the sun; he furnishes the light of light; he has put springs in the dry land and has set stars to minister to the greater lights.[9]

The Celts as well had a sense of the presence of the Trinity, Mary, and the angels attending them in the midst of the common chores of daily living:

> I will kindle my fire this morning
> In the presence of the holy angels of heaven,
> In the presence of Ariel of the loveliest form,
> In the presence of Uriel of the myriad charms.

I lie down this night with God,
And God will lie down with me;
I lie down this night with Christ,
And Christ will lie down with me;
I lie down this night with the Spirit,
And the Spirit will lie down with me;
God and Christ and the Spirit
Be lying down with me.[10]

This sacramental sense of the immanence of God in nature is similar to that found in Hebrew poetry. God is the transcendent source of the beauty, the variety, and the fecundity of nature.

Augustine

Augustine is often placed at the opposite end of the spectrum from Celtic spirituality because of his emphasis on original sin, the fallenness of humanity, and the need for redemption. Matthew Fox, for instance, includes Augustine as one of the "Fall/Redemption" theologians in opposition to the "Creation-centered" theologians, and states that "the cosmos is excluded from Augustine's thought."[11] But this is an oversimplification. Augustine was greatly preoccupied with human sin, and his views on original sin and predestination were too extreme (for example, he thought unbaptized infants went to hell because of original sin). Like Origen, he saw nature and humanity as standing within a framework of gradations of being, and he passed this framework onto later thinkers such as Aquinas and Bonaventure. But nonetheless he had a powerful sense of the beauty and goodness of God's creation. A few citations will make this clear.

A famous passage from Augustine's *Confessions* describes his ascent from the beauties of the mutable world to the permanence of being.

I asked myself why I approved the beauty of bodies, whether celestial or terrestrial. . . . In the course of this inquiry . . . I found the unchangeable and authentic eternity of truth to transcend my mutable mind. And so step by step I ascended from bodies to the soul which perceives through the body, and from there to its inward force, to which bodily senses report external sensations, this being as high as beasts go. From there I ascended to the power of reasoning. . . . This power, which in myself I found to be mutable, . . . withdrew itself from the contradictory swarms of imaginative fantasies, so as to discover the light by which it was flooded. At that point it had no hesitation in declaring that the unchangeable is preferable to the changeable. . . . So, in the flash of a trembling glance, it attained

32

to that which is. At that moment I saw your "invisible nature understood through the things which are made" (Rom. 1:20).[12]

Martin D'Arcy, in an essay on the philosophy of Augustine, comments that this passage is an epitome of all of his thought. The Neoplatonic hierarchy of being is clear, as is the reason behind the hierarchy: the imperishable is preferable to the perishable. The human being can know both, but only with difficulty ascends to the vision of pure being, that which is the source of all the passing splendors of this world.

And yet Augustine did not despise the beauties of this world, for it is through contemplation of these "things which are made" that he arrives at knowledge of the eternal good, God. Sensible things, the Platonists held, were the reflections of the more perfect beauty of the One. Creatures, Augustine says, are coins that bear God's inscription. "Wherever you turn you encounter the footprints of God made upon his work. It is by them that God speaks to you."[13] Like the Celts, he saw creation as resplendent with the beauty of God:

> Ask the loveliness of the earth, ask the loveliness of the sea, ask the loveliness of the wide airy spaces, ask the loveliness of the sky, ask the order of the stars, ask the sun, making daylight with its beams, ask the moon tempering the darkness of the night . . . ask the living things which move in the waters, which tarry on the land, which fly in the air; ask the souls that are hidden, the bodies that are perceptive; . . . ask these things, and they will answer you, Yes, see we are lovely. Their loveliness is their confession. And all these lovely but mutable things, who has made them, but Beauty immutable?[14]

Augustine believed that this beauty would only increase in the world of the resurrection, when the whole creation would become transparent to the presence of God:

> Wherefore it may very well be . . . that we shall in the future world see the material forms of the new heavens and the new earth in such a way that we shall most distinctly recognize God everywhere and governing all things, material as well as spiritual.[15]

The thought of the great African doctor, then, embraces both the idea of a hierarchy of being and the sacramental beauty of the creation, which will only be heightened in the new heaven and the new earth. In spite of his emphasis on hierarchy, he does not imagine that humanity is called to dominate nature. In fact he associates domination with fallen humanity: the lust to dominate is characteristic of sinful communities. H. Paul Santmire comments: ". . . governance . . . is for Augustine a rule not of arbitrary power but of wisdom and propriety, a dominion that allows all things to be and to function in ways appropriate to their natures."[16]

The Medieval Period

St. Francis of Assisi

Francis lived from 1182 to 1226. The son of a wealthy merchant, he experienced a profound conversion at about the age of twenty-four. In a famous confrontation, before the bishop of Assisi, he gave back to his father all that he had, including his clothes, and walked off naked into the snow to serve his new Lord, Jesus Christ. Later, he gathered a small group of followers around him at the church of Portiuncla. This became the germ for his new religious order, the Franciscans, which grew very rapidly. Francis's short life was marked by an intense personal love of Jesus, including mystical visions of the Lord, and by extreme austerities, but also by a great love of nature. Sometimes he is thought of as a kind of nature mystic or flower child, but his intense personal love of Jesus and his austerities put him in a different class altogether.

There is no better expression of Francis's thought than his well-known "Canticle to the Sun," one of the few writings from his own hand that has come down to us. Following is an excerpt:

> Most High, all-powerful, good Lord,
> Yours are the praises, the glory, the honor, and all blessing.
> To You alone, Most High, do they belong,
> And no man is worthy to mention Your name.
> Praised be You, my Lord, with all Your creatures,
> Especially Sir Brother Sun,
> Who is the day and through whom You give us light.
> And he is beautiful and radiant with great splendor;
> And bears a likeness of You, Most High One.
> Praised be You, my Lord, through[17] Sister Moon and the stars,
> In heaven You formed them clear and precious and beautiful.
> Praised be You, my Lord, through Brother Wind,
> And through the air, cloudy and serene, and every kind of weather
> Through which You give sustenance to Your creatures.
> Praised be You, my Lord, through Sister Water,
> Which is very useful and humble and precious and chaste.
> Praised be You, my Lord, through Brother Fire,
> Through whom You light the night.
> . . . .

Praise and bless my Lord and give Him thanks
And serve Him with great humility.[18]

Francis combines in his devotion two qualities: an intense personal love of Jesus and a powerful vision of the whole creation reflecting God. This latter is typically medieval and is rooted in the sacramental sense of the church. We shall find it also in St. Thomas Aquinas (a Dominican) and in Francis's follower, St. Bonaventure (a Franciscan).

St. Thomas Aquinas

One of the developments in the 800 years separating Augustine from Thomas Aquinas was a new appreciation of the laws by which nature acts.[19] Whereas Augustine thought of God as acting directly through nature, so that in a sense everything in nature is a miracle, Aquinas sees God as sustaining a process of natural causes so that both God and nature act to bring about an event. God is the "primary cause" because he sustains everything in being; nature is the "secondary cause" because, by its own laws and operations, it brings about events. "God," Aquinas assures us, "carries out the order of His Providence through the intermediacy of lower causes."[20] The same effect, therefore, comes from both God and from a natural agent.[21] Now this conception of nature, foreign to the Bible and Augustine, is an essential first step in the development of natural science. For if God acts directly to bring about events, no science of nature is possible, for there are no natural laws and operations to study; everything is done directly by God.

At the same time, however, Aquinas retained the notion of hierarchical levels of reality, based on Neoplatonism. This he explained in more technical language than did Augustine. God, he says, is the fullness of being ("being" means the act or energy of existing); creatures *participate* in being. By "participate" Aquinas does not mean that creatures are little pieces of God.[22] God is being by essence. Creatures are created, meaning their being has been given to them (and is maintained) by God. God is their cause; they are God's effects. Being, therefore, which is infinite in God, is partial in creatures. It is not theirs by essence, but they receive it from the Creator. Another way of putting this is that God is the infinite act or power or "energy" of existing. Creatures possess their own act or energy of existing, which is given to them and sustained by God. Because God's essence is to exist, God cannot come into being or pass away, whereas creatures do.

God, therefore, as infinite being, transcends creatures infinitely. But at the same time, God, as the efficient and sustaining cause of every being, is innermost in every being, closer to them than they are to themselves. Aquinas, therefore, retains both the fullness of God's transcendence and the fullness of God's immanence, which must be held in complementarity if we are to have an adequate understanding of God.

Now this means that there is a commonality in being between God and creatures. Creatures are not little pieces of God (as pantheism would hold). But they are not wholly other than God either. Each creature has its own act of existing, yet is sustained by God, who holds all things in existence. There is thus a similarity between creatures and God, but also a dissimilarity (see the section on analogy in the next chapter). We can imagine this through the use of a metaphor. Suppose that there is a cosmic library containing every book ever written and every book that could be written. This would have to be an infinite library. I have one book of my own. There is an exact copy of it in the cosmic library, of course, but yet it is my book. The library has another copy. In the same way, God's being is infinite; my limited being is created, but is similar to God. Like all metaphors, this is imperfect, but perhaps conveys some idea of the relation of creatures and God.

This structure of participation—a partial sharing by creatures in a quality or "perfection" (Aquinas's own term), which is infinite in God—is reflected as well in other qualities: life, for example. Minerals have no life, plants have some, animals more, humans still more, and God is the fullness of life, communicating it to all creatures by degrees.[23] And the same is true of other qualities such as beauty, goodness, activity, self-possession, and so on.

Aquinas's notion of hierarchy therefore is based on the idea of inclusion, not exclusion. Higher creatures contain all the capacities or powers of lower creatures, but additional capacities besides. Thus minerals have the bare power of existence; plants have this, but also the power of life and growth; animals have these, but also sensation and locomotion; humans have these, but also intelligence. Similarly, higher creatures have more of any good, say life or activity, than lower creatures. So in its essential aspect, Aquinas's conception of the hierarchy is not based on domination, power, or will, but on greater and greater degrees of participation in a perfection, such as life, activity, mind, which is infinite in God, but limited in creatures.[24]

Aquinas therefore does not think that lesser creatures are degraded or of no value. On the contrary, they are good in themselves. Thus he declares, in answering the question why God made such a diversity of kinds of creatures:

For he brought things into being in order that His goodness might be communicated to creatures . . . and because His goodness could not be

36

adequately represented by one creature alone, He produced many and diverse creatures, that what was wanting to one in the representation of the divine goodness might be supplied by another. For goodness, which in God is simple and uniform, in creatures is manifold and divided; and hence the whole universe together participates [in] the divine goodness more perfectly, and represents it better, than any single creature [even humans!] whatever.[25]

All creatures together, then, reflect God more perfectly than any single creature (including humans). As with Francis, the whole creation is a kind of sacrament reflecting God. And yet Aquinas also declares that in a sense the purpose of natural creation is to serve human beings:

In the order of the universe, lower beings realize their last end chiefly by their subordination to higher beings. . . . We conclude, then, that lifeless beings exist for living beings, plants for animals, and the latter for men. . . . In a certain sense, therefore, we may say that the whole of corporeal nature exists for man.[26]

This is underscored by Aquinas's teaching that, in the resurrection, most of nature (plants, animals) will not survive; only the four elements (fire, air, water, earth), human beings, and angels will grace the world of the resurrection. This seems inconsistent with his statement (above) that the universe is enriched by the presence of many and diverse kinds of beings.

In sum, Aquinas's thought represents an important and indispensable step toward the development of natural science: the distinction of properly natural (or secondary) causes from the primary cause, which is God. But his ideas on the hierarchy of creation are ambiguous. On the one hand, the hierarchy of being is based on inclusivity and participation in the infinite being of God, and creatures are valuable in their own right, contributing to the sacramental representation of God. On the other hand, they exist to serve human beings and do not survive in the resurrection.

St. Bonaventure

St. Bonaventure (1217–1274) was a Franciscan, and so stood in the tradition of Augustine and Neoplatonism rather than in the tradition of Aristotle, as did Aquinas. Accordingly, he was influenced by Augustine's sacramental vision of nature, but also by the vision of the founding patron of his order, St. Francis, who thought of fellow creatures (and addressed them) as brothers and sisters. Bonaventure's theology of creation was rooted in the idea of the overflowing of the goodness of God.[27] From this fecundity the Father generated the Son, and from this fecundity flowed also the whole created order. Zachary Hayes writes: "The God of fecund love is compared [by Bonaventure] with a vast and living fountain. Flowing from that fountain as something willed and loved by God is the immense river of creation. The world of nature is a vast expression of a loving will."[28] More than Aquinas, Bonaventure sees the whole of nature as a mirror reflecting God:

> The *greatness* of things also—looking at their vast extension, latitude and profundity, at the immense power extending itself in the diffusion of light . . . clearly portrays the *immensity* of the power, wisdom, and goodness of the Triune God. . . . Likewise the *multitude* of things in their generic, specific and individual diversity of substance, form, or figure, and the efficiency which is beyond human estimation, manifestly suggests and shows the immensity of the three above-mentioned attributes of God. The *beauty* of things, too, if we but consider the diversity of light, forms, and colors in elementary, inorganic, and organic bodies, as in heavenly bodies and minerals, in stones and metals, and in plants and animals, clearly proclaims these three attributes of God.[29]

Like St. Thomas, Bonaventure's theology is centered on the hierarchy of being. Even more than Aquinas, he observed that the higher reality (God) was expressed through the lesser, more limited realities (nature). This is a sacramental sense of nature, prominent in the medieval period. One of his greatest works, *Journey of the Mind to God*, is based on the principle of contemplating God in sensory things, and then mounting up to God through the contemplation of higher and higher things. We have seen this contemplative process before in Augustine. But like Aquinas, Bonaventure made little provision for the survival of animals and plants in the resurrection. His sacramental vision of nature was not consistently carried through into the afterlife.

William of Ockham and Ockhamism[30]

A revolutionary view of God and nature emerged in the thought of William of Ockham (c. 1285–1347) and his followers ("Ockhamists"). Against Aquinas and most medieval and ancient thinkers, Ockham argued that universal essences (such as "being" or "human nature") exist only in the human mind and not in nature. "No universal really exists outside the mind in individual substances, nor is any universal a part of the substance or being of individual things."[31] Thus at a stroke Ockham swept away the universal natures upon which medieval metaphysics was erected. There was no universal "being," nor any universal forms or qualities, such as goodness, love, or wisdom, which creatures and God shared. "Being," "goodness," and the like were merely concepts formed by the mind to classify similar characteristics in individuals. This means, then, that there is no "being" in which God, angels, humanity, and nature share; therefore there is no continuity in being between God and creatures.

This undercuts a sacramental view of nature, for the theology of the sacraments depends on a notion of participation in being. Sacraments are visible signs, instituted by Christ, that mediate the grace they signify. For a sacrament to mediate grace, in some sense it must be able to participate in that grace. The eucharistic bread, for example, mediates the presence of Christ precisely because it participates, in a mysterious fashion, in the very being of Christ's risen body. Similarly, if nature does not share in any way in the being of God, if God is "wholly other," it is hard to see how nature can communicate God's presence.

Another consequence was the loss of real participation by human creatures in the being or qualities of God, especially grace. For Aquinas, sanctifying grace was an indwelling accidental form, a *habitus*, that truly sanctified persons, by allowing the charity of God to dwell in them (cf. Rom. 5:5). This was similar to the Eastern idea of *theosis* (deification) except that Aquinas used Aristotelian categories to explain it and the East did not. But both traditions understood grace to be a real participation in the Spirit and charity of God. Such an understanding of grace does not exist in Ockham's system. Ockham insists that God can accept or reject any creature regardless of the created habit or form that dwells in them. "For Ockham grace is not a power which is communicated to man, renews him, and qualifies him for meritorious actions, but God's indulgence, whereby he accepts man or not, as he pleases."[32] Thus God could save a sinner without supplying indwelling grace, or damn a person who possessed indwelling grace. For Aquinas, this would not be possible, because in damning a saint, God would be acting against his own justice and wisdom, and hence against his own nature. But for Ockham, we cannot know God's nature, nor is God bound by any "essential" qualities like justice or wisdom; whatever God wills is just, simply because God wills it. Most Ockhamists argued that God could justly command someone to hate him. The gift of eternal life is entirely due to God's inscrutable will and choice: God is not bound by human ideas of justice, goodness, fairness, or the like.

The effect of Ockham's theology, then, is to replace a hierarchy of *being* with a hierarchy based on *will* or power. God relates to creatures not through a participation in being, but through will and power. Ockham delighted in speculating on the possibilities of God's absolute power; for instance, he argued that God, according to his absolute power, could have chosen to become incarnate in an ass or a stone.[33] For Ockham, then, salvation depended on God's will alone, not on any intrinsic merit of creatures or on their participation in God's being or grace. Grace is not the indwelling love and life of God; it is God's freely given acceptance of the person for eternal life. Ockhamists did argue that God would *probably* save those who did their best, but God was not obliged to: he could damn them.

Ockhamism had enormous effects on European thought. In the first place, it laid the groundwork for naturalism and the emergence of modern science. Ockham had argued that the intellect knows individual entities directly, against Aquinas and others who held that what the intellect knows is the universal form of the species. Thus Ockhamists turned their attention to the understanding of concrete empirical facts rather than abstract metaphysical forms and essences. By emphasizing the contingency of creation rather than necessary metaphysical structures (form/matter, etc.), which could be discovered by reason alone, Ockhamism prepared the way for empirical investigation, for one can only discover the laws of a contingent universe by experimental investigation, not by abstract reasoning.

Second, Ockhamism prepared the way for the Protestant Reformation. Luther was trained by Gabriel Biel, a fifteenth-century Ockhamist, and Luther's thought follows Ockhamism in its rejection of any hierarchy of being. For Luther, human beings relate to God primarily through God's will, not through God's being. Both Luther's and Calvin's view of God and salvation followed Ockhamism, which stressed God's absolute omnipotence and freedom, and that persons are saved by faith alone. Commensurate with this went a weakening of the sacramental principle. Luther and Calvin retained some sacraments, but the center of their theology was the preached word of God, whereas the center of Catholicism and Orthodoxy is sacraments.

Third, Ockhamism emphasized that only individual beings are real, and that the individual is what is known by the intellect, in contrast to the medieval view that the individual was a part of a larger whole. Ockhamism thus led to the development of modern individualism, which was a reaction against medieval emphasis on the corporate, organic unity of the whole people (and the church), as well as a reaction against the unity of humanity and nature. Thus the three movements that most shaped the modern worldview, science, Protestantism, and individualism, all flowed from Ockhamism, which must therefore be accounted the decisive intellectual change from medieval to modern times. By driving a wedge between the being of God and the being of creatures, and exalting God's will over his being, Ockhamism led to the modern conception of God as external to creation and creatures: a God not bound by any intrinsic justice or goodness, whose will is absolute and inscrutable—a God who could damn a saint, who governs more by command than by inviting creatures to participate in his grace. This in turn led to the absolutely sovereign God of Calvin, and eventually to the detached, extrinsic God of Deism.

■ The Seventeenth Century and After

During the seventeenth century, through what has become known as the scientific revolution, a mechanistic view of nature gradually displaced a more organic view of nature—at least among the best educated and scientifically inclined. Over the next three centuries, this mechanistic view was to become the dominant view of the world. Left behind were the Aristotelian and Thomistic approach, with its talk of act and potency, form and matter, formal and final cause, as well as the Neoplatonic approach, represented by Bonaventure, and (in a very different way) by the tradition of Renaissance magic, a vitalist view of nature, epitomized in Paracelsus. The "mechanistic philosophy," as it was called, entailed major changes in the understanding of matter, of causality, and of God. I will discuss these in order.

Both Platonism and Aristotelianism thought of individual things, from dust motes to humans, as composed of matter and form. Matter was not what we now think of as matter: it was a purely amorphous, passive potency, with the capacity to receive form. It was the (substantial) form of a thing that made it the kind of thing it was: a rock, a piece of gold, a rose, a horse, a human being. Form did not mean external shape; it meant the internal organizing structure of a thing. For gold, it would be what makes gold gold and not another metal, like lead. For a plant, a horse, a human, it was tantamount to the soul: the internal principle that made a thing what it was. Any individual existent thing, then, from a rock to a person, was a composite of both form and matter. The presence of the internal form was critical. For inherent in the form was also a goal, an inbuilt end (called the final cause), which directed the entity toward its goal. For an acorn, this would be to become an oak, to flourish, and to leave progeny. For a baby, it would be to become an adult human being. Christian thinkers interpreted the form of the person as the soul, and the final cause of persons to be union with and knowledge of God. Much of Aquinas's theology, both of nature and the human person, was based on the ideas of form and final cause. God directed nature through the ends (final causes) of particular creatures. And what was essential to human beings, what was right or wrong for them, as well as their goals or destinies, was all based on the idea that the essence of the human person was a rational soul informing the matter of the body.

Now this whole metaphysics of nature was displaced, gradually throughout the century, by another metaphysics, that of atomism. The fundamental constituents of matter were not forms and pure potentiality, but hard, massy, tiny atoms, which differed one from another by their shape. From these atoms all physical things were built up. Intrinsic forms and final causes were abandoned. What was left were atoms and the laws of nature. And these laws, as Galileo observed, were written in the language of mathematics. What was important in the new science (as it was called) was to discover and describe the laws of nature in mathematical terms. This was precisely what Galileo did with the law of falling bodies, and Newton did with his laws of motion. The heart of the new scientific method was (1) new abstract concepts, (2) observation and experiments to test hypotheses, and (3) mathematical formulation of laws.[34] Thus Galileo rolled balls down inclined planes to test his hypotheses about the rate of fall, and eventually formulated the law in mathematical equations. But what about the abstract concept? Most people think of science as rooted in common sense observation. In fact, it is rooted in highly abstract concepts, which are far from common experience. Galileo's formulations for the rate of fall only work in a vacuum; under ordinary conditions, air resistance greatly affects the rate of fall. Newton's idea of gravity, as a force acting at a distance, is so far from any common experience that it was ridiculed in Newton's lifetime. The idea that bodies will remain in motion forever until another force intervenes—the law of inertia—is completely contrary to common experience, and never observed on earth.

A corollary of the new conception of matter is that so-called "secondary" qualities—color, taste, smells, and so on—are not properties of atoms or objects, but are experiences of human consciousness. Galileo observed: "I think that tastes, odors, colors, and so on are no more than mere names so far as the object in which we place them is concerned, and that they reside only in the consciousness."[35] What remains are qualities that can be measured: size, weight, momentum, and so on. In this respect, also, the new science was highly abstract. The "real" world was not the world of colors, sensations, and humanly observed beauty; the "real" world was colorless, odorless atomic particles, following abstractly conceived physical laws, like gravity. For reductionists, like Francis Crick, this remains true today: what is real is molecules and their relations, not our emotions, thoughts, consciousness, or experience.

The effect on notions of causality was profound. Swept away were final causes—any sense of an innate goal or teleology in things, or any sense of internal forms, or souls. (The exception here was the case of human beings; most early scientists still kept the idea of an immortal human soul.) In Descartes's philosophy, for example, animals were thought to be merely complicated machines. What remained was what Aristotle had called "efficient cause"—the cause that "makes" something, and "material cause," the matter of which a thing is made. Nature came to be seen as a great machine, obeying mechanical principles; it was regularly compared with a clock. Newton's conception of the planets orbiting around the sun, governed by the force of gravity and inertia, their motions calculable and predictable, like a clock, became the emblem of this view of the universe.

This new "mechanical philosophy" was correlated with an Ockhamist view of God. God was conceived of as the supreme lawgiver, the one who impressed natural laws upon matter, and so held the world in ordered being. But there was no sense of nature participating in and sacramentally expressing higher levels of being. God related to nature extrinsically, through will, not intrinsically, through participating being. This paralleled Calvin's view of God: the supremely sovereign lawgiver, whose will decided the fate of each human atom, who was powerless to oppose or cooperate with the action of grace. God's grace, indeed, for Calvin, was like a force, such as gravity, which the units of humanity, like the units of matter, could not but obey. In an essay arguing the relatedness of the mechanist view of nature and the radical sovereignty of God, Gary Deason observes:

> Thomas interpreted Aristotle's principles inherent in nature (e.g., form and final cause) as powers instilled there by God, which God used in his providential work. God *cooperated* with natural powers in a way that respected their integrity while accomplishing his purposes. When the mechanists rejected Aristotle's understanding of nature, they simultaneously rejected the theory of God's cooperation with nature.[36]

Now this does not mean that early modern scientists were nonbelievers; quite the contrary. Most, in fact were Christians, and many were devout. Galileo, Descartes, and Pierre Gassendi, an early exponent of atomism and mechanism and a Catholic priest, remained faithful Catholics. Johannes Kepler was a Lutheran who was convinced that God had used the principles of geometry in constructing the universe and that he (Kepler), in his astronomical work, was revealing the work of the Creator. Many of the founding members of the Royal Society, the world's first scientific society, were Calvinist Puritans or Anglicans. Robert Boyle, one of the founders of modern chemistry, and a convinced mechanist, was a devout Anglican, who bequeathed money to found the Boyle lectures "for proving the Christian religion." Boyle wrote:

> When with bold telescopes I survey the old and newly discovered stars and planets . . . when with excellent microscopes I discern nature's curious workmanship; when with the help of anatomical knives and the light of chymical furnaces I study the book of nature . . . I find myself exclaiming with the psalmist, How manifold are thy works, O God, in wisdom hast thou made them all![37]

Boyle and many others were convinced that the way to prove God existed was through the exhibition of design in nature. John Ray, a seventeenth-century botanist, wrote a book entitled *The Wisdom of God Manifested in the Works of the Creation* (1691). Isaac Newton also shared this view, and wrote, concerning his *Principia,* "When I wrote my treatise about our system, I had an eye upon such principles as might work with considering men for the belief of a Deity."[38] Newton (though an Arian), in fact wrote far more pages of (unpublished) theology than he did of science, a fact that has recently come to light with the publication of his papers.

This "physico-theology" lasted for well over a century. Indeed the high point of it probably came in the early 1800s, with the publication of William Paley's *Natural Theology* (1802). Paley's account was widely accepted as a convincing argument for the existence of God from nature, right up until Darwin's *Origin of Species* (1859) provided an alternate naturalistic account for the appearance of design in nature.

Thus the immediate effect of the mechanical philosophy of nature was to provide arguments from design against atheism. But the problem was that it relied far too heavily on design, as the single slender thread linking nature to God. At the same time, it severed the unity of God and nature that we find in the Bible and in the sacramental visions of the Celts, Augustine, and Bonaventure, and the commonality of being that we find in Aquinas. Even by the latter decades of the seventeenth century, intellectuals were using the model of nature as a mechanism to argue for Deism, the idea that God had designed the universe and set it in motion, like a clock, but then did not intervene in nature to direct it. For to hold that God had to intervene would be to hold that his design had been imperfect in the first place, a point made by Leibniz and others. The whole idea of intervention fits the image of God having to reach in, interfere, and direct the creation from the outside, rather than guiding it by internal principles and his internal presence. Thus the effect of the mechanical view of the world was to exile God from nature and the universe. When Darwin supplied a more plausible explanation for the apparent design in nature, with his theory of random variation and natural selection, the whole edifice of physico-theology based on design collapsed within decades. And by the twentieth century, when nature had come to be seen as a more or less self-sufficient system, at least by the intellectual community, God had been marginalized from the universe. Stephen Hawking's idea of God, for example, in *A Brief History of Time* (an enormously influential book, which sold over seven million copies), is that of a possible explanation for the beginning of the universe. He has no idea whatsoever of God as immanent, as a necessary context for the universe, or as the source of existence from whom the universe draws its being. Hawking toys with the idea that God may have gotten the universe going, but then proposes a hypothesis that eliminates even the idea of a beginning, and with it (as Carl Sagan points out in his introduction) any need for a creator.

■ Darwin's Revolution

Charles Darwin did not invent the idea of evolution; it had been in the air for about a century. What Darwin supplied was a fully naturalistic explanation for the mechanism by which evolution takes place. His ideas derived from observing animal breeders and from reading Thomas Malthus's *An Essay on the Principle of Population,*[39] which noted that while food increases arithmetically, populations increase exponentially, and that therefore populations always will exceed the limits of their resources. More will be born than can possibly survive. Darwin reasoned therefore that those that survive will be better adapted to their environment than those that perish, that the more fit will leave more progeny, and that therefore every population will tend naturally to a better fit or adaptation to its environment. He called this process "natural selection." Adaptation to the environment, in this scheme, is not due to God, but to the inevitable mechanical workings of nature. Let us consider an example. Robert Boyle noted that God had so arranged it that lambs were born in the spring when there was the most grass and feed for them. Darwin would explain this as follows: any sheep that had lambs born in the fall, or winter, or any time other than the spring, would not leave progeny, because their progeny would die off. Therefore their strains would be eliminated from the breeding population, just as animal breeders culled their herds of all but the choice animals, whom they would breed, so as to maximize a particular trait, such as speed in race horses or milk production in cattle.

Darwin put forth these ideas in his *Origin of Species* in 1859, and he backed his theory up with arguments and a plethora of examples culled from a lifetime of observation of nature. Even now his book makes very convincing reading. Contrary to the stereotypes, though, Darwin's ideas were not immediately accepted by everyone in the biological sciences, nor were they opposed by all Christians. There were significant scientific objections to Darwin's theory (by paleontologists and others), while, on the other hand, some theologians, such as Charles Kingsley, argued that evolution was simply God's way of creating. Eventually, however, Darwin's ideas carried the day. The discovery of modern genetics was a great boost to the theory of evolution, so much so that the combination of Darwin's idea of natural selection and genetic theory is now called "neo-Darwinism." Now, it seems, biologists can hardly resist the temptation to explain everything from physical traits to cultural traits to morals and religion by reference to evolution.

Darwin's ideas were a watershed in the history of religious thought for several reasons. First, he provided a purely natural explanation for the origin of apparent design in nature. William Paley had used the example of the eye, arguing that so intricately constructed an organism, so perfectly adapted to its task, could only come from a Designer. Darwin provided examples of eyes in the animal kingdom in many stages of formation, from mere light sensitive spots to fully developed eyes, and argued that they could have emerged by the process of natural selection. This undercut the whole argument for design on which physico-theology had rested its case. Second, he provided an apparently plausible explanation for the origin of species (though not for the origin of life—he thought that all existing species had derived from a few elemental forms). There was no need to refer to God in the explanation. Third, he changed the view of nature. Where Jesus saw God as caring for all creatures (see Matthew 6), and the Romantics saw nature as suffused with a celestial glory, Darwin saw nature as a struggle for survival, "red in tooth and claw" in Tennyson's famous phrase. As Herbert Spencer put it, it was the "survival of the fittest." This somber picture of nature affected the image of nature's creator. Darwin himself pondered the example of a wasp that stung and anesthetized a caterpillar, keeping it alive so the wasp's young could eat it at their leisure. He could not persuade himself that a good God would have designed such cruelty. As a result, he lost his Christian faith and ended his life as an agnostic. Fourth, Darwin also provided a naturalistic explanation for the emergence of the human species (in *The Descent of Man* [1871]) and so called into question God's creation of the human species and God's creation of the soul.

■ Conclusion

Thus in three millennia we have moved from biblical cultures, through Celtic Christianity, Augustine, Aquinas, and Bonaventure with their sacramental vision of nature, through Ockhamism, through the period of early modern science with its mechanistic view of nature, to the modern period, which sees nature as an autonomous system and God as external to the system. The crucial intellectual change was the cleavage between the being of God and the being of the world. This doomed any sacramental understanding of God's relation to nature, for nature could no longer participate in or fully express God. The sacred cosmos was replaced by a world of dead matter.

What was the fate of the hierarchy of being in all this? We saw that as a result of Ockhamism, the hierarchy of God and creatures came to be understood as a hierarchy of will, not shared being. The advent of the mechanistic view of the universe did not change this, but reinforced it. God's being was utterly different from that of the material universe, and God was rendered extrinsic to the universe, perhaps intervening occasionally. During the eighteenth and nineteenth centuries, thinkers still accepted the idea that human beings had souls and were gifted with reason and freedom, which could not be reduced to mere matter. But by the late twentieth century, the belief in soul was largely lost in intellectual circles. What was really real was matter; the mind was, and is, seen as an emergent property of the material brain, which does not survive bodily death (see chapters 6 and 7). In addition, hierarchy has come to be seen entirely in terms of dominance and oppression. There is a hierarchy of complexity recognized in the sciences, but all qualities are ultimately rooted in matter. Essentially, then, there is no hierarchy, except a hierarchy of material complexity and function.

This is a vision of the world that has scarcely existed before in history, except for a few Greek atomists. The loss of any credible notion of hierarchy means that the universe is no longer seen as stemming from a divine source (*hieros arche* = sacred source). No longer is it "In the beginning was the Word" (John 1), but "In the beginning was matter/energy." Second, the resurrection and afterlife generally are hard to imagine in the modern cosmos. There is no room for anything like resurrected states of matter, or (increasingly) for a soul that survives the death of the body. The loss of hierarchy means that we can't really say that one behavior or state is better than another. If everything is equal, how can we? Science cannot make pronouncements on value, but if naturalism has stripped the world of any transcendent source, it is hard for religion to make convincing value claims either. Finally, without any metaphysical hierarchy, there cannot be a sacramental view of the world and nature, since such a perspective implies that nature mediates a more permanent and glorious reality to us.

In the next chapter, we will consider God and God's relation to nature from the perspective of contemporary science and theology. Later chapters will compare naturalistic and theistic accounts of the origin of the universe, evolution, and human nature, and argue that the theistic account is more cogent.

God and Nature
Unity or Dissociation?

■ In the last chapter we saw that the development of scientific thought was based on the notion of nature as a mechanical system. This notion meant that God was conceived of as extrinsic to nature. To act in nature, God had to intervene from the outside, so to speak. There was no longer an inner connection between God and nature, as there had been in ancient and medieval thought. Furthermore, nature was thought of as mechanical, but God was personal. Hence nature might reflect the design of God, but any deeper unity between God and nature was dissolved. There was thus a dualism between God and nature, which was reflected in the thought of Deism. But even among orthodox Christians, the new mechanistic conception of nature repositioned God as an extrinsic creator. Medieval philosophical categories, such as being, which provided a conceptual link between God and nature, were rejected by the new science and by the Reformers (Luther and Calvin). This conceptual dualism between God and nature has lasted into modern times. The ancient and medieval insight that nature expresses God as in a sacrament survives in pockets (for example, in some Catholic thought), but has no influence in academic discourse. Thus emerged the great modern problem: how can God, who is thought of as detached and distant from the world, relate to the world and act in it? If nature is a mechanical system, following its own internal laws, how can God influence it, except by breaking these laws and interrupting the processes of nature?

Probably the most widespread contemporary strategy for reconciling science and modern theology is what is called "emergentism." This philosophy holds that as matter becomes more complex, especially in the human brain, consciousness, intelligence, intentions, and personality develop or emerge. This means that personal consciousness, intelligence, and our personalities themselves cannot survive the death of the brain, for they are emergent properties that are dependent on the brain. But such an approach, widespread among scientists, philosophers, and many theologians, leaves the problem of the dualism between God and nature unsolved. For, on an emergentist account, how can God have a mind, intelligence, perception, intentions, or even consciousness? For God is not physical or material. So if these properties depend on a complex physical system, then God cannot possess them. Therefore the divorce between God and nature becomes even more acute.

In this chapter I will propose a way of understanding the relationship between God and nature that preserves an inner unity between God and nature, while also holding that there is a great, even an infinite, difference between God and nature. If God were "wholly other," then God could not influence nature, including humans, nor could we influence God. There must then be a similarity and inner unity between God and nature. I will discuss this in terms of being, context, information, beauty, and sacramental presence. Nature, I argue, expresses God as a kind of sacrament. I also will hold that consciousness is not just an emergent property of complex material systems, such as the brain, but is an essential quality of God and spirit. God might be thought of as infinite consciousness; creatures as limited, embodied instances of consciousness. Therefore there is an inner connection between spirit and matter. Matter is, as it were, limited and crystallized spirit, a suggestion made by Catholic theologian Karl Rahner.[1] The best way to make this connection clear is to approach spirit and matter in terms of a unity of information.

To develop this theme, it will first be necessary to discuss two preliminary theological questions. How can we know God at all? How can we speak of God, who transcends all human language and categories?

The Bible represents God as communicating through revelation. God reveals his will and to some extent his nature through prophets, such as Moses, Elijah, Isaiah, and Daniel. For Christians, the supreme revelation comes through Jesus the Christ. But what is revealed through Jesus is not only teachings, but also a whole way of life and being. Jesus' death on the cross reveals the depths of God's love more effectively than any words. The words and deeds of the prophets and of Jesus were recorded in Scripture (filtered through the lens of human interpretation), and passed on by the believing communities.

There remains the problem of how to express God in human language and concepts. The Bible does this through the use of multiple and complementary images. A good example can be found in Genesis 1–2. In Genesis 1 God is portrayed as a cosmic Creator, who surveys the whole of the creation and sees that it is good. But in Genesis 2 God is portrayed as walking in the garden, talking to his creatures like a father. Which image is correct? Theologically, they both are. For God is (in Jewish, Christian, and Islamic thought) the universal, cosmic Creator. But God is also personal, and relates to each human creature in an intensely personal way. So we need both images of God. The best way to describe this mode of portrayal is complementarity. No single image, or theme, or concept is adequate to describe God. Multiple, complementary images are necessary. And so we find God portrayed in the Bible as both a warrior (Exod. 15:1–21) and as love (1 John 4:16), as angry (the wrath of God, Pss. 6, 7, 30, 145) and forgiving (Ps. 145; Matt. 6:14), as a rock (Pss. 18, 28, 42, 62) and as a father (Matt. 6:9) and as a mother (Isa. 49:15; Matt. 23:37). Theologically, God is described as both just and merciful, transcendent and immanent. God is whole, and therefore cannot be described simply as this or that, but must be described by complementary terms and images.

Interestingly enough, we can see traces of this complementarity in the created order. For most things in our creation exist as one pole of a complementarity.[2] Subatomic particles exhibit the properties of both particles and waves. They might carry a right- or left-handed spin, or negative or positive charge (e.g., electrons and protons). Complex molecules come in right- and left-handed forms (so-called isomers). DNA is made up of two complementary strands. Higher animals are male or female. And so on. (Sometimes light and dark or good and evil are thought of as complementary. But complementary entities must exist on the same level. And this is not true for light and dark or good and evil. Darkness is simply the absence of light, not a "thing" in itself. So also with evil, which, in the classic definition of St. Augustine, is an absence of good.)

One of the ways that the principle of complementarity has been manifested is in the use of both positive and negative terms for God. In the West, we use mainly positive terms, such as "God is love" or "God is wise" or positive images such as "God is father." But in the East, God, or absolute reality, is usually portrayed in negative terms. The Upanishads characterize the supreme reality, Brahman, as "not this, not that." Mahayana Buddhists describe the supreme reality as "Sunyata": emptiness or void. Negative terms are not lacking in the Western theological tradition, but are less frequent than positive terms. When we describe God as "infinite" we mean "not finite." Also, we say that God is not material, not bound by time and space, not mortal, and so on. The idea behind using negative images to characterize the divine is that such images do not limit God. The danger of using positive images, such as father, is that we tend to think of God as just like we are, only bigger and more powerful. But such anthropomonphism makes God into a finite being, and so not God. So the East, especially in its philosophical traditions, has chosen the way of negation. The West has chosen principally the way of positive affirmation.

But we need both ways, for they are complementary. Aquinas brings both ways together in his doctrine of analogy. An analogy is a comparison of two things that are similar in one respect, but unlike in others. For example, we speak of strength of muscles, strength of will, strength of an argument. As W. Norris Clarke explains, there is a certain similarity among these uses of the term *strength*, but it is used in a different manner in each case.[3] Aquinas notes that this is true of the positive terms we apply to God. We say that God is love. But God's love is not just like human love. The love of God must be imagined as stripped of all human limitations. God's love is not limited, not selfish (as most human love is), not short-sighted, not partial to one rather than another. God's love is perfectly unselfish, without favorites, tender but also just, and so on. So God's love is similar to human love in one respect, but dissimilar in others. And in its positive aspect, God's love is similar to human love, but infinitely greater. Thus Aquinas attempts to combine both the negative and positive ways of predication. A positive epithet may be used of God, but we must realize that God is *not* like its limitations, and in God, the positive attribute (love, wisdom, justice, consciousness, etc.) is infinite, not limited as in humans.

There is another difference between God's love and human love. God is love essentially; God cannot not love. In God, to love and to exist are identical—God would not be God without love. But humans can love or not love and still remain human. To love and to exist are *not* identical in humans. Therefore the *mode* of love in God differs from its mode in humans. God is love essentially; in humanity love is contingent. This difference is true of all predicates that are applied to God. For example, we say that God exists and that humans exist; rocks, planets, stars exist. But God exists essentially, necessarily. God cannot not exist, whereas humans, rocks, planets, stars, indeed all created things, can not exist. So the mode of existing differs radically between God and any creature (including angels). Therefore the analogy does *not* mean simply that God possesses more of some quality or perfection (Aquinas's term) than creatures do. It is not true that God is infinite existence and we are, say 1 percent existence, as if existence in God and creatures was used in exactly the same way (i.e. univocally) so that God and creatures were on the same scale but God was greater. The whole *mode* of existence differs radically between God and creatures. And so any universal, positive term—love, knowledge, existence, wisdom, activity, etc.—is predicated of God in a different way than it is of creatures; it is predicated *analogically*, not *univocally*.[4] It is therefore proper to speak of a similarity between God and creatures, and of a unity within difference, but not of a continuity.

Another example might make the difference between univocal and analogical predication more clear. We often use the term "know" both of animals and humans. A dog knows its master, or it knows the way home. Humans know things through the senses (as animals do) but can also know things that cannot be directly sensed, but must be inferred and reasoned to, for example, that the earth goes around the sun, that atoms are made up of electrons and protons, that protons are made up of quarks, and so on. Is "knowledge" the same in animals and humans? Certainly humans have more knowledge than do animals, but human knowledge is also qualitatively different. Humans can know a thing through abstract reasoning. So while it is correct to use the term "knowledge" of both animal and human cognition, knowledge in humans is not just quantitatively different but also qualitatively different from knowledge in animals. There is an analogical similarity, then, but not a univocal similarity.

With these theological tools in mind, we can attempt to discuss God, who is the proper subject of theology. Yet it is harder to talk about God than anything in the world, because God is not material and transcends us infinitely. It is easy for us to talk about things that we transcend and can control, like automobiles, rocks, molecules, atoms, and so on. But it is hard to talk about things that are beyond us, and that we cannot control, like the weather, the biosphere, the universe. But even these are easier to approach than God, for they are physical and can be measured and tested. But God cannot be either measured or tested. And so we talk about God in human terms, but must keep in mind that these terms are analogies, that all analogies are limited, and that they therefore need to be balanced by complementary terms.

Western images and concepts of God are mostly personal. One exception might be God as Spirit, but even the Spirit has been personalized. The greatest insight of Christianity is that God is intrinsically relational and therefore personal, that God is love. This insight is expressed in the doctrine of the Trinity: God is persons in relation even "before" the world had been created. God is personal and relational from all eternity. And God's object in creating the world is to draw it up into a sharing of the relation of love that God enjoys within himself. As St. Paul writes: "God's love has been poured into our hearts through the Holy Spirit that has been given to us" (Romans 5:5). The aim of the Christian life is to share in God's love and manifest it to the world. For this, personal relationship to God, through Jesus and the Holy Spirit, is essential. Love takes priority over following the law, as St. Paul makes clear in his letters. By contrast, the typical expression of God in nontrinitarian theistic religions such as Judaism and Islam is law, the expression of God's will. The duty of the believer is obedience, following God's law.

Nevertheless, personal images and concepts of God, which are positive images, are not the whole story. What about the complementary negative way? God is not a person in the same way that we are persons. God is not an individual, as a human person is. God does not have our moods or prejudices, cannot be bribed, does not take sides, does not think like we do. "For as the heavens are higher than the earth, so are my ways higher than your ways and my thoughts than your thoughts" (Isa. 55: 9). The danger of merely personal images of God is that we make God into a larger version of ourselves, complete with our own prejudices. Furthermore, if we conceive of God only personally, we inevitably think of him as extrinsic to nature. But God, as well as being transcendent, is also immanent, nearer to us than we are to ourselves.

Therefore I will sketch out some ways of thinking about God in impersonal or transpersonal terms. These are to be understood as complementary to the use of personal images and analogies. I will consider God as Being, as context, as information, as beauty, and as sacramental presence. These analogies will, I hope, show that God, even though infinitely beyond all creatures, nonetheless retains an inner unity with them as well. This unity in turn has implications for how God can act in the world and in our lives.

■ God as Being

As we have seen in chapter 2, "being" is a major theme in the thought of Aquinas, and in Thomistic thought today. Modern people think of being as simply a static fact—something is, there is no mystery about that, it's just there. Aquinas, however, thought of being as an action, like running. (He used the Latin infinitive *esse* = "to be" to denote what we translate as "being"). The "first act" of any creature is its act of existing, which comes from God, who is the source of all be-ing. God is be-ing by essence; creatures are be-ing by participation (as Aquinas puts it).[5] He writes, "To participate is to receive as it were a part; and therefore when anything receives in a particular manner that which belongs to another in a universal or total manner, it is said to participate."[6] God, then, is Being in all its infinite fullness. Creatures are partial, limited acts of being.

Aquinas argues that God is the absolute act of existing; God cannot not exist. He is therefore not "a being." For if God were *a* being, his being would be limited, as it is in creatures, and we should have to ask, "Who created God?" But God is not the sort of being who *can* be created. God is infinite and uncreated; creatures are finite and dependent. Every creature, including angels, receives its act of existing as a gift from God. All are held in being by God. If God withdrew this support, creatures and creation would vanish.

God's holding of creatures in being is part of God's act of creating, as Aquinas explains in the following passage where he asks if God is present in all things.

> God is in all things, not, indeed, as part of their essence, nor as an accident, but as an agent is present to that upon which it works. . . . Now since God is very being by His own essence, created being must be his proper effect; as to ignite is the proper effect of fire. Now God causes this effect in things not only when they first begin to be, but as long as they are preserved in being; as light is caused in the air by the sun as long as the air remains illuminated. Therefore as long as a thing has being, God must be present to it, according to its mode of being. But being is innermost (*L. intimum*) in each thing. . . . Hence it must be that God is in all things, and innermostly (*L. intime*).[7]

The relationship of creation, then, means first that creatures are not part of God; God transcends them infinitely, as an agent transcends what he makes. But God is also present to them most intimately (*L. intime*), actively holding each creature, and all of nature, in being. Furthermore, God's act of creation did not occur just once; it is ongoing, occurring in every instant. Being or existence, then, is not just a static fact; it is radically dependent on God's ongoing creative activity. Because of the fact of creation, God is not a stranger to the natural world but is actively united with it.

Another way to clarify the active nature of being is to use a modern analogy: energy, which everyone thinks of as active. It might help illuminate Aquinas's original conception of "being," which people today think of as passive and static. In the analogy of energy, God is infinite energy and the source of all energy. God brings into existence creatures, who are limited, specified, embodied parcels of matter/energy, held in existence by God's infinite energy. We must resist, however, the temptation to think of creatures as small capsules of God's own energy. (This is the doctrine of pantheism.) Creatures are not little bits of God; they are separate beings with their own matter/energy, and subject to the laws of nature. Therefore, while there is a similarity between the energy of God and the matter/energy of creatures. There is also a discontinuity between God and creatures.

For Aquinas, God is simple. Simplicity means God does not have parts, or even aspects (except as perceived by us). God is the fullness of being, but also of goodness, activity, wisdom, etc. And in God, being and goodness are not different qualities; they are one. To an extent this is true of created beings also. To the extent that a being exists, it is good, and it is active. But, at least for Aquinas, there are degrees of goodness and activity. A stone exhibits a minimal degree of activity—it exists. Plants exhibit more, animals more yet, humans even more. This is Aquinas's notion of the hierarchy of being (which is also a hierarchy of life, goodness, self-possession, knowledge, and so on).[8] It is not based on oppression; it is based on inclusion. Higher levels include all the capacities of lower levels, but more besides. Thus stones can exist, but not much more. Plants can exist, but also grow and reproduce. Animals can grow and reproduce, but also move about and sense things. Humans can do all these but also reason and know God. Humans are superior to animals, and animals to plants, then, in the sense that they possess all the capacities of those beings lower than they, but additional capacities besides. But God possesses all good qualities, especially the quality of existing or being, infinitely. Because each type of creature reflects only a very limited aspect of God, the fullness of God's goodness and beauty is best represented by the whole array of creatures and creation.

This idea of participation in qualities that are infinite in God but limited in creatures also underlies the Catholic and Orthodox understanding of grace. Grace is the help given by God to humans. Usually it is a participation in God's Spirit, life, and love that elevates human capacities and allows us to exceed our natural abilities. A classic biblical expression of this is Romans 5:5: "God's love has been poured into our hearts through the Holy Spirit that has been given to us." Grace is therefore a real sharing in the love of God, which elevates and transforms human love, so that we can love those whom we could not love naturally (for example, our enemies).

This, then, is a traditional way to think of the similarity and unity between God and creatures. But it is based on a premodern science of qualities and does not speak to a scientific mentality, which demands more quantified and measurable models. Therefore I will consider other ways, God as context, God as the communicator of information, God as beauty, and God as sacramental presence.

■ God as Context

A second way to conceive God is as the ultimate context of the universe. Everything that is familiar to us exists in some wider context. We exist in a society, which exists within the larger biosphere. The earth itself exists within the context of the sun and the solar system. The solar system exists within the Milky Way galaxy, which in turn exists within a supergalactic system. And many supergalactic systems exist within the observable universe. This is one kind of context—the physical context. We also however exist within social, historical, and cultural contexts, which influence our individual behavior. If one is a naturalist, all there is to context is physical environment, history, and culture. But if one believes in spiritual reality, then we also exist within an unseen spiritual context, which, for traditional Jews, Christians, and Muslims, includes angels and God, the ultimate context. Other religions would describe different spiritual contexts. Asian religions think of karma as a spiritual force that determines human destiny. It is almost a definition of religion that some sort of nonphysical context—God, Brahman, the Tao, Sunyata, ancestor or nature spirits—affects human life.

Now we usually think of contexts as passive. But in all the examples given above they actively affect the behavior of individual entities. Consider some examples. Sixty-five million years ago the earth was devastated by the impact of an asteroid or comet that was ten miles in diameter. The blast caused by this impact resulted in continent-wide forest fires, a huge pall of dust and smoke in the air, and the effective blockage of the sun for months, possibly for years. This in turn caused the death of much of the plant life, and the animals that lived on those plants—the dinosaurs—died out. Now one could argue that this was not the context at work, but one part affecting another—the asteroid impacting the earth. But the orbit of the asteroid was caused by the gravitational attraction of the sun and its attendant planets by other asteroids, and perhaps partly by adjacent stars. So it was the whole system or context that caused the impact. The same is true for any large-scale weather changes that affect continents and epochs.

Physicist John Polkinghorne gives another striking example. He notes that in the air in a room, each molecule will have, on the average, fifty collisions with other molecules in one ten-billionth (10^{-10}) of a second. Let's say we want to predict what direction a molecule will be going after the fiftieth collision. It is easy to work this out on paper; we simply calculate the angle of the initial collision, then the angle of the second collision, and so on, just as we would calculate the directions of billiard balls in a chain collision. The problem is that in our calculations we will have to approximate the angle of each collision, say, to three or four decimal places. But such an approximation is not nearly enough, because any uncertainty in the approximation increases exponentially as we go from collision to collision. So how accurately would we have to know the initial angle of collision of the first molecule to be able to predict the direction of the last molecule in the chain? The answer is amazing:

We will make a serious error in our prediction if we have failed to take into account an electron (the smallest particle) at the edge of the observable universe (the furthest distance away) interacting with the air molecules in the room through gravity (the weakest of the fundamental forces in nature). So even as simple a system as air in so short a time cannot have its detailed behavior worked out without literally universal knowledge of all that is happening.[9]

This is an illustration of what is called "chaos theory." Tiny changes in the context can affect the behavior of events far removed in time and space. The usual example concerns weather. It is said that a butterfly flapping its wings in Beijing will affect the weather in Minneapolis two weeks later because, as in a chain collision of pool balls or air molecules, discrepancies increase exponentially. If the initial angle of collision is changed ever so slightly, that change is amplified with each successive collision, so that by the fiftieth collision, the direction of the last ball or molecule is unpredictable.

Ordinarily, of course, we ignore such tiny forces as the gravitational pull of an electron, or even a star, in another galaxy. They will not affect us as we drive to work or eat dinner. But tiny, imperceptible causes, such as solar winds and sun spots, do affect sensitive systems, such as weather, and occasionally even computers. Context is important because it can exert causality.

"Contextual causality" can be understood by examining artifacts as well as examples from nature. Complex artifacts, such as computers or automobiles, are usually explained by the interaction of their parts. But their design is also shaped, within limits, by their context. A modern automobile is built for a planet with a certain gravity, a solid surface, and oxygen in the atmosphere, for a culture with paved roads, petroleum, metallurgy, plastics, and electronics, for intelligent beings of a certain size, with two arms and legs, eyesight, etc. Imagine a modern automobile transported to a very different planet, or back to antiquity, and one can immediately appreciate the role of context in its design.

Virtually any artifact is designed to fit within a narrow range of contexts. Indeed, it is a routine method in archaeology to reconstruct the context from an artifact.

At the level of genes, cells, and organisms, the principle of contextual causality is also present. The scenario of genetic reductionism popularized by Richard Dawkins, Edward Wilson, and others is oversimplified. Steven Rose, professor of biology at the Open University in England (and a persistent critic of Dawkins), argues that the organism itself influences which of its genes are expressed: "It is the organism in interaction with its environment . . . that determines which of its available genes are to be active at any one time."[10] As an example he cites the work of Monod and Jacob, who found in their experiments that "bacteria which did not normally possess the enzymes required to metabolize the sugar lactose would synthesize them if their food supply was restricted only to lactose."[11] Another example of the context influencing gene expression is found in Mississippi alligators. According to biologist Brian Goodwin

> it turns out that what determines whether a developing alligator will be male or female is the temperature it experiences during a critical period of its embryonic development. Eggs that develop in the temperature range 26°–30° C are all female, while between 34° and 36° they are all male. In between 31° and 33°, the switch occurs and eggs developing at these temperatures can be either sex.[12]

Similar mechanisms are found in some species of lizards, turtles, and fish.

One of the most striking examples of contextual causation is found in natural selection itself. The principle here is that it is the environment that influences which individuals of a population will survive and leave progeny, and hence, it is the environment that shapes the character of the population itself. In *The Beak of the Finch*, Jonathan Weiner describes the change in the shape of the beaks of finches on the Galápagos over a period of decades due to seasons of drought.[13] Natural selection is the most favored contemporary biological explanation for the development of species, and probably constitutes the clearest case of contextual causality.

Finally, we can turn to the relation of the individual person and human culture. No one doubts that there is a large cultural factor in all individual behavior. If I am building a house, a whole series of cultural factors enter into that activity: government laws and building codes, financial markets and funding opportunities, availability of land, materials, tools, skilled labor, the technology developed over generations and passed onto the builder, and so on. In building, I am constrained to use the land, materials, tools, techniques, and funding methods that are available in the culture. On one level it is I who build the house; on another level, it is an expression of the culture.

These examples point to the phenomenon of "nested contexts." Contexts are rarely simple; they are generally complex, and can be broken into levels, at least conceptually. The immediate context of a gene is the genome and its cell; the cell in turn is embedded in an organism; the organism in a local environment, which is embedded in a regional environment, which broadens out to a whole ecosystem, to a planet, to a solar system, etc. In the case of the builder, there is an immediate local environment, then a wider regional environment (which might set building codes), then a wider national and global environment (which might affect the financial markets, for example).

Contexts, then, affect individual processes. And if God is the ultimate context of the universe and human life, we might also expect that God would affect both nature and human life as well. The question is: How does God do this? By what mechanism can God affect natural processes?

The question of how God acts in the world is one of the major problems facing the theology-science dialogue. A number of answers have been proposed, which we will take up in a later chapter. Briefly, some of the principal suggestions are these: (1) God affects events at the quantum level. Since quantum events are to some extent indeterminate, they allow a space for God to act without breaking natural laws. God can determine the indeterminacies at the quantum level, and so affect the course of events—for example, evolution—at the macroscopic level. (2) God acts by the input of "active information" (John Polkinghorne's proposal). This could occur at the quantum level, but perhaps at other levels also. (3) Arthur Peacocke has proposed that God acts through what he calls "whole-part influence." He gives examples from chemistry and biology of the whole system influencing the action of the parts.[14]

■ God Acts through Information

One way that God acts in nature and human life is through the input of information. God can be thought of as containing an infinite field of information, and creatures as forms specified by information. Information can be thought of as specifying one of two (or more) alternatives. Imagine a switch that can be on or off. If the switch is turned on (or off), that is one bit of information. This is the basis of digital computers. They operate on a binary code, in which enormous sequences of switches are specified as 0 or 1. Binary code can generate, among other things, a number series. Our system of numbers is a code based on ten units (0–9), but other number systems are possible, including one based on two units (0, 1). In such a system the code 1 would equal our number 1. But 10 would equal our 2, 11 would equal 3, 100 = 4, 101 = 5, 110 = 6, and so on.

By specifying a "1" or "0" for each place in a binary number, information can then specify a number or numbers. In the same way, it can also specify patterns. A circle is defined as a figure that includes all points at a specified distance from a center. Thus, if you place a point at the center of a piece of graph paper, and trace out all the points three inches from that center, you will draw a circle with a three-inch radius. And if you specify that a circle six inches in diameter be drawn on a piece of paper, you are specifying one particular pattern or figure out of an infinite number of possible patterns that could be drawn on the paper. All kinds of figures and curves can be generated by mapping out the points that fit certain equations. This is how computers generate intricate curves and patterns. They merely trace out all the points that correspond to a certain formula. Information, then, specifies not only numbers but also patterns, or, to use an older philosophical term, form. Information specifies form.[15] A simple example of this is a blueprint, which specifies the form of a house or building.

We can think of God as containing an infinite field of information. In the mind of God exist all possible beings, events, and universes, including those that have not been created but could have been. When God creates, some of these possibilities become actualized—a universe is created with its own immanent laws. In Scripture, God is portrayed as creating through his word (Gen. 1:3; John 1:1–3), the Logos, which became incarnate in Christ (John 1). Of course God did not physically speak and produce sound waves. But the point is that the word conveys information. God's act of creating, as depicted in Genesis 1–11, is an act of forming the waters of chaos into an ordered world.[16] And this involves the input of specifying information (information specifies form). Exactly the same could be said of the creation of our universe; it is created with highly specific laws and physical constants that allow it to develop as it has. Almost any other set of laws and constants would have resulted in a universe that could not support life (see next chapter).

If God is the infinite fullness and source of information, any creature can be thought of as embodying specified information, a discrete parcel of information. It is information that specifies the form of the creatures. Take a hydrogen atom as an example. It is formed of one proton and one electron, and is located in a specific place and time. No other atom can occupy the same space it does. The electron is a particle/wave characterized by a very specific mass, negative charge, a certain spin, and a certain orbit in the atom. The proton likewise has a specific mass, charge, spin, and position. So the form, place, and time of the atom are described by specifying information. But this information is embodied in a particular formed parcel of matter/energy, constituting this particular atom. The atom, like every material creature, is a discrete, specific, embodied instance of information. The information contained in the hydrogen atom is much easier to describe than that contained in a human being, but the principle is the same in both cases.

Notice that we do not have to say that the information embodied in the creature—say, an atom—is a portion of the information that exists in God. The creature is not a little part of God, for Christianity asserts that the Creator is radically different from the creature. But it is a peculiarity of information that it can be copied. Two copies of the same book contain the same information, yet individually they are distinct. The same is the case with creatures and God. A creature may embody information that is also in the mind of God, but the creature is its own reality, not a little fragment of God.

Now this way of thinking about God and nature as information, like any analogy, is limited. Most important, it does not capture God's personal quality. Also, information in our world is always carried by a material vehicle, but the same information can be carried by many different vehicles. The message "S.O.S." can be sent by Morse Code; it can be written on a banner; it can be called by voice; it can be put in different languages; it could even be conveyed by bodily gestures. But it is the same information. So the nature of information itself is peculiar. It needs to be carried by a material vehicle (in our world) but in itself seems to be non-material.

Information, as used here, is similar to mathematics, for mathematics conveys information and describes the form of things. If we say that the dimensions of a room are 12 feet by 15 feet by 10 feet high, we have conveyed the shape of the room, though not the materials of which it is made. The fundamental physical laws structuring our universe are expressed scientifically in mathematical terms. The law of gravity, for instance, is expressed mathematically as f (force) = G (a constant) x mass 1 x mass 2 divided by the square of the distance between the masses. With this equation (and others), and a knowledge of the masses and momentum of the sun and the planets, one can compute the trajectory of the planets moving around the sun. So mathematics expresses information. And sciences such as physics, and chemistry, and to some extent biology, depend on mathematics. But is mathematics itself material? Some mathematicians think of numbers and mathematical entities as simply concepts that exist in our minds. Others think of them as real universals, like Platonic forms: numbers somehow exist apart from our minds, but are not material. In either case, though, numbers and mathematical entities are not material. In this they are like information.

A Christian way of understanding "where" mathematical entities exist is to understand them as existing in the mind of God. They then have a real objective existence outside of our minds, but at the same time are not material. But this is equivalent to saying that all possible informational patterns exist in the mind of God. Some of those have been actualized in our created universe, and perhaps some have been actualized in other universes; we cannot know this, however.

Many philosophers and scientists now recognize that the fundamental reality of our universe is matter/energy plus information. For matter/energy, as we encounter it, always has specific forms, such as a hydrogen atom. And these forms are specified by information. Some thinkers posit that information (or patterns) is the fundamental constituent of reality. Edward Fredkin has argued such a position.[17] This also seems to be the basis of Stephen Wolfram's book *A New Kind of Science*. Ray Kurzweiler holds that the fundamental reality in our universe is patterns: "In my view, the fundamental reality in the world is not stuff, but patterns."[18]

My own position is that physical reality is energy plus form or embodied information. These are two different principles. Our world is not just patterns on some supercomputer, as Fredkin seems to think. Information is always embodied in matter/energy. As far as the form or pattern goes, any two hydrogen atoms are identical. What differentiates them is the matter/energy of which they are composed. It is a characteristic of matter/energy that it must occupy one position in space at a particular time. That is what makes one atom of hydrogen different from another. In Aquinas's terminology, matter "individuates" form—it makes a given form into this individual thing, not that individual thing. There are then two principles in any physical thing: matter/energy and form, or embodied information. This is a modern way of understanding what Aristotelians and medieval philosophers called matter and form.

But information seems analogically similar in both creatures and God. Creatures are specific, embodied parcels of information; God contains an infinite field of information and is the originating source of information and form in the world. There is, then, a similarity between God and the world in the area of information. One way that God interacts with the world is through the input of information. This, indeed, seems to be a fundamental aspect of God's creation. But God's act of creation is not only in the past; it continues today. And as in the beginning, so now, God's creativity involves informing the world through the input of information.

It is important to insist again that understanding God as information is an analogy. It is only one way of understanding God, and needs to be complemented by the use of personal analogies. By itself, the analogy of information obscures God's qualities as person, God's responsiveness, and God's relationship of love to his creatures. But at least the analogy of information, like that of being and context, provides a way of bridging the gap between God and nature, and of understanding how God might act in nature.

◼ A Sacramental View of God and Nature

A sacrament makes God present through the mediation of a physical sign or ritual. It is more than a memorial, but less than direct seeing. Roman Catholics, Eastern Orthodox believers, and Anglicans have held a "high" view of sacraments. For these groups, sacraments are not merely memorials. Rather, through them, God and Jesus are really made present to the believer. The Eucharist is the best example. The bread and wine really become the body and blood of Christ; they are not merely symbols. Luther agreed with this position. He insisted that the bread and wine really conveyed the body and blood of Christ, which were given "in, with, and under" the forms of bread and wine. Luther, however, rejected transubstantiation. He thought that Christ was really present in the bread and wine, but the bread and wine did not cease to be bread and wine (as the Catholics taught). Other Protestant groups took somewhat less realistic views of the sacraments. Joannes Oecolampadius, one of Luther's interlocutors at the Colloquy of Marburg, insisted that the Eucharist was a mere symbol. When Christ said, "This is my body," he meant it metaphorically, just as when he said, "I am the vine" (John 15:5).

I will take a realistic view of sacraments here, and argue that in such a view, nature itself can be seen as a kind of sacrament, for it also makes God present. This is not the only way to see God, but it is a way of overcoming the separation of God and nature, and God and everyday life, which is so prevalent in Western society.

How is God made present in a sacrament? Consider the Eucharist. Traditional Catholic thought focused on an objective change in the bread and wine—their substance became the substance of the body and blood of Christ. Recent Catholic thought has taken a subjective turn. In the liturgical action, the bread and wine take on a different meaning to the believers; they are "transignified," and so become the body and blood of Christ. I would argue that there is both a subjective change in the believers and an objective change in the bread and wine: the latter become incorporated into God, much as food becomes incorporated into the body. But in the case of the eucharistic bread and wine, the incorporation is a metaphysical reality, not visible in any change in the bread and wine themselves. The bread and wine remain what they are, but are incorporated into a greater whole, the risen body of Christ, and so become part of a larger substance.[19]

Something like this is also true of nature. Nature exists in God.[20] It is similar in its being, its information, its design, and its majesty, and its seeming infinity to its Creator. It has its own autonomy and its own laws. Yet, from a Christian perspective, it exists within a larger context: God. And, to the eyes of faith, it mediates the presence of God to the beholder. But it does not perfectly reveal God (as it will be in the resurrected state). It is also partly opaque. Life in nature cannot exist without death and suffering. In this sense, nature is an imperfect manifestation of the divine.

Another way to understand the sacramental character of nature is to think of it manifesting the presence of God through its beauty. As the psalmist declares: "The heavens are telling the glory of God; and the firmament proclaims his handiwork" (Ps. 19:1). This sacramental sense is what we find in Celtic spirituality and in the thought of Augustine (chapter 2). For him, the loveliness of the earth, sea, and sky proclaim the greater loveliness of Beauty immutable, who made them all. In a famous passage in the *Confessions*, he exclaims: "Late have I loved thee, O Thou Beauty both so ancient and so new, late have I loved Thee!"[21]

Modern people think of beauty as subjective—in the eye of the beholder. But ancient and medieval people thought of it as an objective quality in things themselves. Beauty was the harmony of diverse parts in an ordered whole. Today we would say it is unified diversity. Think of a beautiful painting or a beautiful musical composition. There is diversity—a monochrome painting would not be beautiful, nor would a musical composition of only one note. But there is also unity. Without it there would be cacophony, or mere randomness, but no composition. Admittedly these qualities are hard to define. That is one reason that the classical understanding of beauty was displaced by the modern subjective understanding. But to define beauty as entirely subjective is to say that we have, literally, no standard as to what counts as art (or music) and what does not. This subjectivity is apparent in some contemporary art. There is no canon or standard of beauty against which it is measured, so virtually anything can count as art.

Curiously enough, beauty is also an important standard in science. Physicists especially speak of the beauty of a theory. By this they mean mathematical beauty, concision, and elegance. Einstein's equation $E = mc^2$ is a very concise statement that relates energy to mass and the speed of light in one simple formula. Fundamental and diverse physical principles are united harmoniously into a simple equation. It is beautiful. Some physicists, such as Paul Dirac and Murray Gell-Mann, have held that beauty is a better test of a successful physical theory than almost anything else. Even if the initial experiments do not confirm the theory, if the theory is beautiful, they hold, it will eventually be proven true. Beauty, in this sense, is hard to define precisely (and even harder to explain to students), but physicists are confident that they can recognize it. It is not just a subjective quality; it is a quality in the physical theory itself.

There is more to beauty than just unified diversity. Ancient authors such as Plotinus also spoke of "radiance." Beautiful things seem to radiate a splendor that transcends them. But radiance is precisely the sacramental quality. Beautiful things mediate the presence of a greater reality that is beyond them. And they draw us to that greater reality, whether we call it "God," "Mystery," "Spirit," or whatever.

In both classical and modern theology, God is the fullness of beauty. God is triune: three united in One, and so is the primal image of Beauty. In God all things hold together. All things created, and all things that could have been created, exist first in the mind of God. God not only sees them all, God sees all their possible interrelations, in time and in eternity.

The union of all things with God, through Jesus Christ, is the Christian vision of the ultimate goal and end of nature and human life. For the return of creatures to God and their union with God is Paradise. The mystery of God's will, says the author of Ephesians, is "to gather up all things in him [Christ], things in heaven and things on earth" (Eph. 1:10). There are some magnificent images of this in Dante's great poem, the *Divine Comedy*. In the *Paradiso*, Dante, who is the pilgrim in the poem, sees all the redeemed gathered in the form of a great white rose, with the angels ministering to them, in heaven (canto 31). In the final canto (33), he sees "all the scattered leaves of the universe bound by love into one volume."

Nature, then, is a reflection, an image, of the greater beauty of God. It is not just an arbitrary sign. It really shares in and expresses the qualities that exist in an infinite degree in its Creator. And in this it is like a sacrament; it makes God present, and therefore we can speak of it as a sacred cosmos, for whatever mediates God's presence is sacred. This sacramental quality of nature has been beautifully portrayed in painting, especially paintings of the early Romantic period, such as the landscapes of Frederick Church and the Hudson River school.

But nature is also imperfect. It includes death and suffering, which do not exist in God. Death is a sign of the transitoriness of nature and natural beauty, a frequent theme of poets (e.g., Shakespeare in his sonnets). And because of its imperfection, it points beyond itself to its greater Source. "For now we see in a mirror, dimly, but then we will see face to face" (1 Cor. 13:12).

To be aware of God's presence in nature, we have to attune ourselves to it, through prayer, meditation, contemplation. We have to learn to see nature differently, through the eyes of faith, rather than as an object to be engineered and mined for resources. It is of course sometimes necessary to regard nature as an object. It is necessary to get work done, and it is necessary in scientific investigation. But we can learn to balance this viewpoint with a sacramental understanding of nature, which would bring us more fully into the presence of God in our everyday life.

▓ Conclusion

Different denominational traditions tend to emphasize different ways of conceiving God and God's relation to nature and everyday life. Evangelical traditions stress a personal relationship with Jesus. Catholic spirituality has emphasized more often a sacramental approach—seeing God in all things, as St. Ignatius of Loyola put it. But the personal relationship with Jesus has not been absent from the Catholic tradition, and is especially prominent in the lives of the saints, such as St. Francis of Assisi and Ignatius himself. The Catholic philosophical tradition, especially that of St. Thomas, has understood God as Being. And the scientific tradition, in such figures as Boyle, Newton, Kepler, and others, has understood God as the cosmic designer.

Somewhat the same typology can be applied to Jesus. For evangelical and Pentecostal Christians, Jesus is a personal friend, a brother, the Lord. For the Christians of the Middle Ages, Jesus was seen as the judge, who summoned all humanity before him in the last judgment. But Jesus can also be seen as the incarnate Logos, the Word through whom all things were made, the cosmic designer who created the universe and in whom it all comes together (Eph. 1:10).

If I am right that the great modern problem is that God has become remote from nature and from everyday life, then we need to find ways to open ourselves more to God. For of course it is we who have changed, not God. God remains closer to us than we are to ourselves, and innermost in all things. But we have lost this awareness. One way to bridge this chasm is to discover a deeper personal relationship with Jesus. Another way is to recover the sacramental sense of God's presence in nature. This is an antidote to modern naturalism, in which our horizon is constricted to physical nature alone. These ways are not mutually exclusive. The greatest Christians, such as Augustine and St. Francis, have loved God in both ways. The fullness of Christianity, and the practice of that fullness in everyday life, requires that we be open to all authentic understandings of God and Jesus.

In the following chapters, we shall take up the question of how well naturalism actually explains the world and humanity. We shall consider naturalistic and Christian explanations for the origin of the universe (chapter 4), evolution (chapter 5), and human nature (chapters 6 and 7).

4

Origins
Creation and Big Bang

■ Even children want to know the answer to the question "Where did the universe come from?" Traditionally, the answer was: God created the universe. More recently, the answer sometimes is the "Big Bang," as if the Big Bang had replaced God. But the Big Bang theory provides no explanation for either the origin or the existence of the universe. It describes the evolution of the universe from about the first 10^{-43} seconds on, but gives no explanation for what actually caused the initial explosion. Now, if everything can be explained by reference to nature, as naturalism claims, then one would think that naturalism would be able to explain the origin of nature itself, that is, of the universe. If it could, the universe would not require any explanation, reason, or cause outside of itself. And in fact a number of scientific naturalists have put forth some highly speculative theories that claim to show that the universe, in effect, created itself and requires no further explanation. The most famous of these is that of Stephen Hawking. Yet when one examines these theories closely, one finds that they do not in fact provide any explanation for the existence of the universe. Either they claim the universe originated from some preexisting condition, which itself requires an explanation, or (as in Hawking's case) they think that if the universe has no datable beginning, it therefore needs no explanation. The first strategy merely pushes the crucial question one step further back; the second confuses it with another question (When was the beginning of the universe?) and so avoids it altogether. In fact naturalism has no explanation for the origin and existence of the universe. I will maintain, however, that theism does have an answer to this ultimate question. Let us begin by first considering what the Big Bang theory says, then look at the various theories of origins. After that we will consider why the universe is so finely structured that it can produce intelligent life, and then consider what answers theology might be able to contribute to these questions.

The Big Bang theory is actually quite recent; its roots go back only to about 1930. Before that, the prevalent scientific opinion was that the universe had always existed. This was also the view of Aristotle and Aristotelian physics, and, incidentally, of most Asian religions. (Hindus and Buddhists find the idea of a universe that had a beginning almost incomprehensible. Their view is that the universe moves through vast cycles of creation, growth, decline, and destruction, but is itself eternal; these vast cycles will never end.) Aristotle did not teach a cyclical universe, but nevertheless envisioned a universe that had neither beginning nor end; it had existed from all eternity. Thomas Aquinas, in confronting this Aristotelian cosmology, concluded that human reason by itself—what we would now call science—could not prove that the universe either was eternal or was created. That the universe had been created by God, he held, could be known only through revelation.

During the nineteenth century, the static universe of Newtonian mechanics was gradually replaced with the idea of a universe that had evolved from previous states, but which was still usually thought of as having always existed. Because of the limitations of nineteenth-century astronomical observation, however, the visible universe was thought to be confined to the Milky Way galaxy. It was not until 1923 that Edwin Hubble, working with the new 100-inch reflector telescope at Mount Wilson Observatory, was able to resolve the Andromeda galaxy into separate stars, thereby proving that it was a galaxy similar to our own. By 1929 Hubble's observations of distant galaxies had led him to the conclusion that most of the galaxies were traveling away from one another at great speeds, and that the farther away the galaxies were, the faster they were receding from our galaxy and from each other. It was this discovery that led to the development of the Big Bang theory. By 1930 it had become widely accepted in astronomical circles that the universe was expanding. This led logically to the idea that the universe might have had a beginning in an initial explosion.[1]

Probably the first proposal that could be called a Big Bang theory (though the term itself was not coined until later) was that of a Belgian Catholic priest, Georges Lemaitre. Lemaitre's idea was that the universe originated in an enormous "primeval atom" that exploded and led to the expanding universe. He was able to demonstrate that Einstein's general theory of relativity entailed an expanding universe. Lemaitre's model, later modified by George Gamow, was the prototype of contemporary Big Bang cosmology.

Still, throughout the 1930s, '40s, and '50s, and into the '60s, Big Bang models of the universe were opposed by the so-called "steady state theory." This held that the universe had always been expanding from eternity, and that new matter was continuously being created as space expanded, so that the density and distribution of matter in the universe remained approximately the same. Interestingly enough, many of the early defenders of the steady state theory were attracted to it partly because it offered support for their *theistic* beliefs. American physicist Robert Millikan, in his 1930 address as retiring president of the American Association for the Advancement of Science, emphasized that the hypothesis of hydrogen being created out of radiation pointed to "the Creator [being] continually on his job."[2] Yet later defenders of the steady state theory, notably the English astronomer Fred Hoyle, an atheist, thought the Big Bang alternative looked too much like a creation from nothing (which pointed to a Creator), and so argued for the steady state theory.

Both the steady state theory and the Big Bang theory envision an expanding universe. In the steady state hypothesis the universe had no beginning but has always been expanding (and always will). The Big Bang theory, by contrast, envisions an expanding universe that seems to have had a beginning. The evidence that decisively favored the Big Bang theory was the discovery of cosmic background radiation in 1965. In that year, American radio astronomers Arno Penzias and Robert Wilson detected very faint radiation in the microwave range: 7.35 centimeters; the radiation corresponds to black body radiation at a temperature of 3.5 degrees above absolute zero.[3] It was soon realized that this was "relic radiation" left over from the Big Bang itself. That is, the radiation that had not condensed into matter after the Big Bang continued to saturate the universe, but expanded along with the universe and cooled, so that it is now about 1,000 times cooler than it was originally. The steady state theory could not account for this ubiquitous radiation, which comes at the earth uniformly from all directions. On the other hand, it was a striking confirmation of the Big Bang theory. Measurements taken by the COBE (Cosmic Background Explorer) satellite revealed in 1992 that there are very slight "ripples" or nonuniformities in this microwave radiation, which correspond to the initial clustering of matter into supergalaxies about 300,000 years after the first moment of the Big Bang. Again, this was a striking confirmation of the Big Bang theory.

■ The Big Bang Theory

What exactly does the Big Bang theory say about the origins and evolution of the universe? Scientists can describe the early moments of the universe back to the first one-hundredth of a second or so, and they can speculate back to about 10^{-43} second (so-called Planck time), but beyond that the laws of physics break down (because of the extreme heat and pressure), and the physical state of the universe cannot be described. There are many speculations about the origin of the energy that was released in the Big Bang, but the theory itself endorses none of these speculations. It simply attempts to describe the evolution of matter from the earliest milliseconds on.

What was the course of this evolution?[4] At Time = 0 seconds (T = 0), presumably the universe would have consisted of a so-called "singularity" in which all the energy of the universe was concentrated in a point of infinite density, pressure, and temperature. But the laws of physics cannot describe such a condition. Just after T = 0, the universe would have expanded and cooled enough so that basic subatomic particles—protons, neutrons, and electrons—began to exist as free particles, along with the photons and neutrinos that formed the bulk of the matter in the very early universe. As the universe continued to expand, it also cooled; like a gas, as it doubled in size, the temperature fell by half. After about three and one-half minutes, when the temperature was about a billion degrees Centigrade, protons and neutrons were able to combine into simple atomic nuclei: hydrogen, deuterium, and helium. The Big Bang theory correctly predicts that the proportion of hydrogen to helium should be about 75 percent to 25 percent.

For the next three hundred thousand years or so, the universe continued to expand and cool, but was still too hot for electrons to combine with the nuclei of hydrogen and helium permanently to form stable atoms. Any hydrogen or helium atoms that formed temporarily would be immediately broken apart by the intense radiation pressure. But once the universe had cooled to about 3,000 degrees Kelvin (i.e., 3,000 degrees above absolute zero), the radiation pressure decreased enough to allow atoms of hydrogen and helium to form and persist. This in turn allowed the photons or radiation to expand freely without being impeded by the hot gas of electrons. As cosmologists say, the universe became "transparent" to radiation. This radiation then would have been at about 3,000 degrees Kelvin; now, since the universe has expanded and cooled by a factor of 1,000, it is observed as the cosmic background radiation whose mean temperature is about 3 degrees Kelvin.

As the universe continued to expand, gradually the hot gases were drawn together by gravity in various regions, beginning the formation of galaxies and stars. Due to the enormous gravitational force crushing the matter together in stars, the interior of the stars heated up to the point where nuclear reactions began. Hydrogen atoms were synthesized into helium (by the same process that fuels the hydrogen bomb), thereby releasing immense amounts of heat energy. This energy, generated in the center of the stars, counteracted the process of gravitational attraction, so that the stars became stabilized and would continue to "burn" hydrogen into helium and give off a steady supply of heat and light for billions of years. Other nuclear reactions also take place in the stars: helium is built up into carbon atoms by nuclear synthesis (three helium atoms combine into one carbon atom, again releasing heat), and other heavier elements are built up. But no atoms heavier than iron can be formed in this way. Eventually, however, very large stars exhaust their nuclear fuel and collapse suddenly (in a few seconds) after burning steadily for millions of years. They explode in supernovae, events in which a single star may give off more light and heat than an entire galaxy of one hundred million stars. Supernovae in our galaxy may be visible even in daylight, as was the great supernova which formed the Crab nebula in 1054 A.D., recorded by the Chinese astronomers. In the process of supernova explosions, neutrinos combine with lighter elements to produce all the heavy elements up to uranium. These heavy elements are ejected into interstellar space, where they form a dust that may eventually recombine (due to gravity) into planets like the earth. Indeed, if this had not happened, human life on earth would be impossible, for it depends on light elements, such as carbon, nitrogen, and oxygen, and also on heavy elements, such as zinc and iodine.

Actually, there is some uncertainty concerning the exact process by which galaxies and stars formed and the process by which our solar system formed.[5] The solar system may have condensed from hot gases circulating around the sun (the hot origin theory), or it may have formed from dust that was accumulated by the force of gravity into asteroids and planets (the cold origin theory). But the main outlines of the Big Bang theory have gained acceptance by virtually all scientific cosmologists. There are no defenders of the steady state theory today. This is not to say that the consensus may not change at some future time, but if it does, it will have to incorporate the evidence that so forcefully points to the Big Bang theory: the evidence that the universe is expanding, the presence of the microwave cosmic background radiation, the ratio of hydrogen to helium and of photons to matter in the universe, and so on.

Thus the Big Bang theory nicely explains the development and evolution of the universe from the earliest times. But it leaves unanswered the big question: What was the origin of the Big Bang itself? Where did all the energy/matter[6] come from? What caused the initial explosion? Scientists are extremely uncomfortable postulating an event without a cause, especially the most massive and explosive event of all time. Inevitably, some very speculative theories have been advanced to explain the origin of the Big Bang. We will survey some of them here.

One idea is that while the Big Bang might represent the beginning of *our* universe, it may not be an absolute beginning. Perhaps there were previous universes that expanded and collapsed and rebounded in another expansion, thus forming an infinite series of "successor universes." This theory is known as the "oscillating universe theory." It is attractive because it avoids the problem of an absolute, inexplicable beginning and precipitates us back into an eternal universe, one that oscillates or pulsates in a cyclical series of expansions and collapses and reexpansions. However, the physical evidence does not favor this theory. Stephen Weinberg comments:

> Some cosmologists are philosophically attracted to the oscillating model, especially because, like the steady-state model, it nicely avoids the problem of Genesis. It does, however, face one severe theoretical difficulty. In each cycle, the ration of photons to nuclear particles . . . is slightly increased by a kind of friction (known as "bulk viscosity") as the universe expands and contracts. As far as we know, the universe would then start each new cycle with a new, slightly larger ration of photons to nuclear particles. Right now this ration is large, but not infinite, so it is hard to see how the universe could have previously experienced an infinite number of cycles.[7]

The question of whether our universe will recollapse depends on whether there is so much matter in the universe that gravity will eventually (billions of years from now) slow the rate of expansion to a stop, then reverse it, so that the universe would implode, collapse back onto itself, and sink into a massive black hole. This question depends on another question: What is the so-called "dark matter" in the universe, and is there enough of it to reverse the expansion? Astronomers know that much of matter in galaxies (perhaps 90 percent) cannot be seen. This can be calculated because galaxies rotate, and at their present rate of rotation they should fly apart, like mud flying off a rotating wheel. The mass of the stars that can be observed in the galaxies simply is not enough to account for the force of gravitation that holds them together. Therefore there must be a large amount of "dark matter" that accounts for the gravitation which holds the galaxies together. But no one knows what this dark matter is. It could be neutrinos. Presently, neutrinos are thought to have no mass (like photons), but if they do have mass, even a very tiny mass, there are so many of them that their cumulative mass might be enough to account for the dark matter. So the question is unresolved.

But recent measurements have shown that in fact the expansion of the universe seems to be speeding up, not slowing down, as one would expect. No one knows how to account for this. But if these findings hold true, then our universe will expand forever, and not collapse.[8] If so, then to defend the oscillating theory one would have to argue that our universe was the last of a long cycle of expanding and collapsing universes, and then explain why previous universes had collapsed and reexpanded, but ours will not. The more likely explanation is that ours is the only universe.

Another recent theory, advanced by Alan Guth, Paul Davies, and others, is that the universe is the result of a kind of "quantum fluctuation," and so is self-creating, the "ultimate free lunch" as Alan Guth puts it.[9] Guth's theory is roughly as follows. Physicists believe that in so-called "empty" space, quantum particles may be coming into existence and vanishing back into nothingness all the time. This can happen because "empty" space is not really empty; it is permeated by energy fields (for example, gravitational fields). In any energy field, subatomic particles may be condensing out of the energy for extremely small fractions of time, and then be dissolved back into energy. These are so-called "quantum fluctuations." Guth has coupled this notion of quantum fluctuations with the idea of inflation, of which he was one of the founders. Guth's inflation theory is that the universe, about 10^{-35} second after T = 0, underwent an enormously rapid period of expansion—much faster than the speed of light—in which its size increased from something like 10^{-52} meters to almost a meter.[10] This period of inflation took place in less than a trillionth of a second. The cause of the inflation is, according to Guth, a supercooled "false vacuum," which created a negative gravitational force, which in turn caused the extremely rapid expansion.[11]

There are some fifty different versions of the inflationary theory. One of the bizarre consequences of Guth's version is that the false vacuum would produce not one but perhaps an infinite number of universes, each separated from one another like bubbles in foam. Another consequence is that what we can observe of our universe would be only a tiny fraction of the total volume of our universe (which is one bubble universe among countless others).

Guth hypothesizes that what causes each of these myriad universes to come into existence is a quantum fluctuation in the false vacuum, before the period of inflation. These extremely tiny quantum fluctuations are then enormously expanded by inflation so that each of them becomes an entire universe. Thus our universe, and others as well, is the result of a quantum fluctuation in the false vacuum, a superdense state that preceded the existence of our universe.

Of course this raises the question, What caused the false vacuum to come into existence, or is it eternal? Guth believes that even the false vacuum must have had a beginning. He guesses that perhaps it too could have arisen as a quantum fluctuation from a state that he describes as "nothing": "One can imagine that the universe started in the totally empty geometry—absolute nothingness—and then made a quantum tunneling transition to a nonempty state."[12] But he admits that even in this state, the laws of physics would still have had to exist, and their origin is a mystery.[13]

Inflation theories, though not directly testable, do explain many problems in the cosmology of the very early universe, such as why the matter in the universe is so evenly distributed, why the temperature of the background radiation is so even, and why the amount of matter in the universe is just enough so that the universe would expand for 12–15 billion years without recollapsing due to gravity, or expand so fast that stars and galaxies could not form. Guth's theory that there are an infinite number of bubble universes seems a good deal more speculative, since there is no way to test it. And his idea that the false vacuum arose from a quantum fluctuation in a geometrically empty universe, that is, a universe without any spatial points, is simply a mathematical conjuring trick, which mistakes a geometrical or mathematical notion of nothing for ontological nothingness (that is, absolute nothingness). A geometrically empty universe is still not "nothing" in the ontological sense. For a quantum fluctuation and quantum tunneling to occur, some sort of energy potential must also exist, and the laws of physics must also exist. And Guth's theory, as he himself admits, gives no explanation for the existence of these. Thus his theory, clever and speculative as it is, merely pushes the problem of the origin of the Big Bang one step farther back. The question still remains: What caused the universe (or the false vacuum) to come into existence, and why do the laws of physics have the form they do, which allows life to exist? At present, physics cannot answer these questions, nor can any other science.

Another author who claims that the universe sprung into being from a quantum fluctuation is the Oxford chemist Peter Atkins. Atkins's surprising contention is that not only did the universe come into being from nothing, it actually *is* nothing: "the seemingly something is actually elegantly reorganized nothing, and . . . the net content of the universe is . . . nothing."[14] Atkins's reason for holding this is that the universe really consists of a balance of opposite energies and forces. The matter/energy in the universe is positive energy, but it is balanced by gravitation, which is a negative energy, so the net energy content of the universe is zero. "Nothing," then, for Atkins, as Keith Ward notes, "turns out to consist of an equilibrium of an infinite number of equal and opposite forces, charges, and so on."[15] Within this "nothing" that is the universe, there can be tiny, chance fluctuations, and one such fluctuation, in a primordial quantum field, could have given rise to our universe. Now the problem here is with what Atkins calls "nothing." Again, it is not absolute or ontological nothing, but a mathematical kind of nothing, like the sum of +1 and −1. His whole case depends on the confusion of these two kinds of nothing. Imagine a person who owns a house, automobile, and personal property, but who also owes creditors a total that equals the value of what he owns. Mathematically, his net worth is zero. But we would hardly say that therefore all of his goods—his house, car, property, etc.—are "nothing," that is, they don't exist! But this is exactly what Atkins is saying about the universe. Because its net energy content is zero, it is really nothing, and therefore can come into existence from nothing. Indeed, there is nothing to be explained!

But clearly, because the forces of the universe balance out, that does not mean that nothing exists. And if the universe arose from a quantum fluctuation of some previous state, something would have had to exist in that state, certainly the laws of physics and some kind of quantum fields. So the explanation that the universe arose from a quantum fluctuation is really no explanation. It is first of all entirely hypothetical and almost certainly untestable, because any evidence of other universes is not available to us. And second, it merely pushes the question of origin one step further back, so that we have to ask: where did the previous universe come from?

Finally, we should consider the theory advanced by Stephen Hawking at the end of his book *A Brief History of Time*. Hawking thinks that when the universe was less than 10^{-45} seconds old, the temperatures were so high and the energy so dense that the laws of physics would have had to be described by a quantum theory of gravity. In this case, he speculates, time itself might become indeterminate, as quantum events are. This would mean that there would be no definable boundary to space-time, and no datable beginning to the universe. Time near the beginning would fuzz out in quantum indeterminacy. Hawking concludes: "The universe would be completely self-contained and not altered by anything outside itself. It would neither be created nor destroyed. It would just BE."[16]

What are we to make of this? In the first place, Hawking's theory is completely speculative, unsupported by any evidence, as he himself admits. But even if it turned out to be true, would the fact that time becomes indeterminate near the beginning of the universe mean that the universe is ontologically independent, without any ontological cause, that is, that it just sprang into existence by itself? Hawking's theory assumes that the laws of physics—the laws governing quantum mechanics, gravitations, etc.—are already in place. Where did they come from? And where did the energy/matter that emerged at the Big Bang come from? Hawking's theory leaves these questions unanswered, and therefore does not give us an explanation of the origins of the universe.

Thus scientific theories of origins, though some of them claim to have explained the origin of the universe, typically assume the prior existence of an ordered whole—some kind of energy field that is structured by the immanent laws of physics. They therefore assume what they wish to explain, that is, an ordered universe, but do not provide an explanation of its origins.

Thus, naturalism has no answer for the biggest question of all: Where did the universe come from?

▓ The Anthropic Principle

There is another question raised by the existence of our universe, namely, why is it structured so precisely as to lead to the existence of intelligent life (the so-called "anthropic principle")? For the order of our universe is quite amazing, and is something of a mystery. It turns out that the physical laws and constants of any universe have to be precisely balanced for complex systems, including life, to exist at all. Not just any universe will produce life; only one in which the physical laws and constants are just right will do. We thus live in a "finely tuned" universe, a fact that has only emerged as the evolution of the universe has come to be more fully understood. Some of the evidence for this follows.

All of the activity and change in nature is due to four fundamental forces: gravity, the weak nuclear force, the electromagnetic force, and the strong nuclear force.[17] These differ greatly in both character and magnitude. Gravity, the attractive force between particles of matter and fields of energy, is by far the weakest of these forces. True, the gravitational attraction between a boulder and the earth is large, but only because the mass of the earth is so large; the gravitational attraction between two boulders sitting in a field, or floating in outer space, is negligible. However, the range of gravity appears to be universal. The gravitational field created by our sun extends throughout the solar system, the Milky Way galaxy, and even the universe. But since gravitational force between two objects decreases proportionally to the square of the distance between them (so that if the distance is doubled, the force decreases four times), the effect of the sun's gravity outside our solar system is vanishingly small. The weak nuclear force, stronger than gravity but still very weak, is associated with neutrinos and the tra?.... tations in subatomic particles. Its range extends to only about 10^{-1b} centimeter from its source. The electromagnetic force produces lightning, magnetic fields, electricity, and the attraction between particles of opposite charge, such as electrons and protons. It is about 10^{39} times stronger than gravity, but like gravity its range is universal and falls off according to the inverse square law. Unlike gravity, however, the electromagnetic force is not cumulative; in most material objects, the positive and negative electrical charges cancel each other out, so that the bodies are electrically neutral. Only rarely do electrical charges build up, as in the case of a lightning storm. Large bodies like the sun and the earth do create magnetic fields, but these are weak. Finally, the strong nuclear force is the force that binds together the protons and neutrons in the nucleus of an atom. It is a very strong force, much stronger than the electromagnetic force, but its radius of action is very small, extending hardly farther than the nucleus of the atom. This is the force that produces atomic energy, including the heat and light of the sun, which are produced by nuclear fusion.

Now we have seen that atoms heavier than helium are built up in the interior of stars over billions of years. Gravity draws together the huge masses of hydrogen and helium, which because of the gravitational pressure heat up and begin the process of nuclear fusion (due to the strong force). The elements built up in the stars are held together by a balance of the strong nuclear force and the electromagnetic force. At the end of their lifetimes, the more massive stars explode in supernovae, in the process creating all the elements heavier than iron through the interaction of neutrinos with other elements (due to the action of the weak force). The explosion also expels these elements into space, where they can condense into planets (due to gravity), which can produce life.

This, incidentally, is why the universe must be so old and so large if it is to support intelligent life. The slow process of building up the chemical elements in the stellar furnaces takes billions of years. The first generation of stars must reach the end of their lifetimes and explode in supernovae in order to create the heavier elements, some of which are necessary for human life. Then the remains of these stars must condense into planets, and a long process of evolution occurs (on at least one planet) for intelligent life to occur. This whole process could hardly take less than ten billion years. Meanwhile the universe has been expanding all that time at enormous speeds. And so, by the time intelligent life has had time to evolve, the universe is already more than ten billion years old, and more than ten billion light years across.

Now it turns out that for this whole process to occur, the four forces must have precisely the character and magnitude that they do. Any slight change in the force of gravity, the nuclear forces, or the electromagnetic force would mean that the universe would not evolve to the complexity required to support life. The laws and physical constants of the universe must be "fine tuned" to an incredible extent for the universe to give birth to intelligent life. According to Paul Davies, if the strength of either gravity or the electromagnetic force were changed by as little as one part in 10^{40}, it would be catastrophic for stars like the sun.[18] Similarly, if the strong force were increased or decreased by only a few percent, the evolution of complex elements inside the stars could not occur. The weak force, also, has to be set within very narrow limits for the build-up of the elements to occur.[19] Stephen Hawking gives a striking example. In the early instants of the Big Bang, if gravity were only a tiny bit stronger, the cosmos would have collapsed long before intelligent life could have evolved. But if it were only slightly weaker, the cosmos would have expanded so fast that the hydrogen and helium would not have condensed into stars and galaxies. How accurately did these two factors—the initial rate of expansion and the force of gravity—have to be balanced? The answer is: within one part in 10^{60}, an incredibly tiny possibility. To use the simile of Fred Hoyle, it is like a tornado going through an aircraft junkyard and assembling a 747 ready to fly. Now it is true that if there was an early period of inflation in the universe, according to the theories of Alan Guth and others, this huge coincidence could be avoided. But inflation theories themselves require a universe fine tuned within one part in 10^{50}.[20]

An interesting story concerning the "fine tuning" required for the universe comes from the English astronomer Fred Hoyle. Hoyle discovered that the nucleus of the carbon atom possesses a "resonance" at precisely the right value so that three helium nuclei will stick together to form one carbon nucleus in the process of nucleosynthesis occurring in the stars. If the resonance level had been 4 percent less, essentially no carbon would have been produced in the stars; had it been 0.5 percent higher, virtually all the carbon produced would have been converted into oxygen. A slight variation in either direction would have meant that virtually no carbon would have been produced in the stars. And since all life is based on carbon chemistry, there would have been no life in the universe.

Now Hoyle had long been an outspoken atheist. As noted above, he had argued for the steady state theory instead of the Big Bang theory in the 1950s because the Big Bang looked too much like a creation from nothing. But the discovery of this resonance level in the carbon nucleus, at precisely the right value to allow the production of carbon atoms, and hence life, shook Hoyle's atheism so much that he concluded that the universe must have been the product of a super intellect. The following paragraph was written by Hoyle for the Cal Tech alumni magazine in 1981:

> Would you not say to yourself: "Some supercalculating intellect must have designed the properties of the carbon atom, otherwise the chances of my finding such an atom through the blind forces of nature would be utterly minuscule." Of course you would. . . . A common sense interpretation of the facts suggests that a superintellect has monkeyed with physics, as well as with chemistry and biology, and that there are no blind forces worth seeking about in nature. The numbers one calculates from the facts seem to me so overwhelming as to put this conclusion almost beyond question.[21]

Now the question which follows from the "fine tuning" of the universe is: What caused it? Was it coincidence? Did the universe have to be this way, for some reason that we don't know? Or was it designed this way by a superintelligent Creator? Let us consider these possibilities. We could argue that there is some necessary reason that the universe must be the way it is, and that it could be no other way. If we could find such a reason, that would explain the fine tuning. But, as Keith Ward observes, "The physical cosmos does not seem to be necessary. We can think of many alternatives to it."[22] Any of the basic forces, or any of the physical constants, could have been slightly different. Some kind of universe would have resulted, probably one much more simple and less complex than this one, such as a universe made up of only hydrogen and helium; but not a universe that could support life.

Another possibility is that there are many, many universes that exist, perhaps quintillions, and that the physical laws in each of them are different. We just happen to live in the one that can support life. To us, that looks like one chance in 10^{60}, but if there are 10^{60} other universes out there, then it is an even chance. If there is an infinite number of universes, each with differing laws, then it would be inevitable that one at least would be fine tuned so as to support life. The problem with this scenario is that there is simply no evidence for these other universes. And it is hard to see how we could ever know if they exist or not, since by definition they are not a part of our universe, and so could never be observed by us. Thus this scenario is an appeal to ignorance. And even if all these trillions of other universes existed, we would still have to explain where they came from. And why are all their physical laws different one from another? Like the quantum fluctuation hypothesis, this scenario just pushes the question one step farther back, into an area that cannot be investigated, and so does not seem to be a satisfying explanation at all.

Thus scientific theories, though they provide convincing explanations for how the universe developed, provide no real explanations either for the origin of the universe, why it exists, or why it is as intricately ordered and "fine tuned" as it is. Science raises these questions but cannot conclusively answer them. This is a point made by some scientists themselves, such as George Ellis, a physicist and cosmologist. Ellis writes:

> If we look at the Anthropic Issue from a purely scientific basis, we end up without any resolution, basically because science attains reasonable certainty by limiting itself to restricted aspects of reality. . . . By itself, it cannot choose between these options.[23]

The third possibility is that the universe was designed by a super-intelligence, to use Fred Hoyle's phrase. (This is also George Ellis's conclusion, in the book cited above.) This leads to a consideration of theological explanations for the origin and the fine-tuned structure of the universe.

■ Theological Accounts of the Origins of the Universe

Traditionally, Judaism, Christianity, and Islam have believed that the universe was created by God. In Genesis 1 God creates the world by his word (and perhaps by his Spirit, which broods over the waters). The Hebrew word *bara*, translated as "create," is only used of Yahweh. But what does "create" mean? In the first place, it means to order. John McKenzie writes: "Even in Gen 1:1 we are probably not dealing with creation from nothing . . . but with God's ordering of chaos into a fixed universe."[24] Second, it means to sustain. Again, McKenzie: "Yahweh sustains and defends the material universe against the forces of disintegration; without the uninterrupted exercise of his saving power, the world will relapse into chaos."[25]

The idea of creation from nothing emerges explicitly only in later Hebrew writings, especially in 2 Maccabees 7:28 (which is not included in the Jewish scriptural canon or in the Protestant Bible, but which is included in Roman Catholic Bibles as inspired Scripture). There, a Jewish mother says to her son who is about to be executed by Antiochus, "I beg you, my child, to look at the heaven and the earth . . . and recognize that God did not make them out of things that existed." It was this belief that became the traditional doctrine of later Judaism, Christianity, and Islam.

What are the implications of creation from nothing (*creatio ex nihilo*)? First, as Aquinas puts it, it means "to produce a thing into being according to its entire substance."[26] This differs from any material change or human artifice, in which there is some preexisting material that is restructured in the change. To build a house, to paint a painting, or even to give birth is not to create from nothing, but to make something new from previously existing elements. (Perhaps the closest we can come to creation is in the realm of ideas: we have a really new idea, and publish it. But even here the new idea will have been made up of other, previously learned, ideas.) Second, Aquinas tells us, it means

> that non-being is prior to being in the thing which is said to be created. This is not a priority of time or of duration, such that what did not exist before does exist later, but a priority of nature, so that, if the created thing were left to itself, it would not exist, because it only has its being from the causality of the higher cause.[27]

So, in creation, the whole substance of a thing is produced from nothing, and what is created depends entirely for its being on a higher cause, without which it would cease to be. Creation, then, comes wholly from God; it does not depend on any preexisting material. While this is not made explicit in Genesis 1 (Gen. 1:2 says, "the earth was a formless void and darkness covered the face of the deep"), I would argue that it is implicit. For what is total formlessness? It is, precisely, nothing. Every existing thing, even empty space, is structured by dimensions, has form, and follows physical laws. Nothing can exist without some form, and so absolute formlessness is nothing. (See chapter 3, note 15.) The ancient Hebrew authors could not express this conceptually, because they lacked the requisite philosophical language, but the closest they could get to it was the poetic image of a formless void, watery chaos, in which all form has been dissolved.

Now creation from nothing is not the same as having a beginning in time. Something that is created owes its entire existence to another source, which is prior to it not necessarily in time but ontologically, that is, in being. Aquinas argues that even a universe that has always existed, an eternal universe, would require a creator as the source of its being and its laws, even though it had no beginning in time.[28] Thus even in the case of an eternal universe, we could ask, Why does the universe exist at all? Why does it have these particular laws and this particular structure? Why not another set of laws and another structure? Hawking's mistake, discussed above, is to assume that if the universe has no beginning in time, it therefore need not have been created. But, as we have seen, this does not explain why the universe exists or why it should have the physical laws and structure that it does, which just happen to be so precisely balanced as to allow life to emerge.

Of course, what is created from nothing may also have a beginning in time. Indeed, that would seem fitting. But it is not essential to the idea of creation from nothing. The question of the absolute dependence of the universe on God and the question of its beginning are really separate questions. Some authors, such as John Polkinghorne, think that the doctrine of creation concerns only the ontological dependence of the universe on God, and that the question of the beginning of the universe is irrelevant to the doctrine of creation. But this separates the ontological origin of the universe from its physical origin, and leaves the latter question—which is a huge question indeed—unanswered. Furthermore, the contention that the universe depends on God, even though its physical origin can be explained by some other cause, is not going to be persuasive to many people. In any case, if one asked, "What would the creation of the universe from nothing look like to a physicist?" the answer is "It would look like the Big Bang!" In present physical theory, energy, matter, space, and time are all interrelated. One does not exist without the others; all would have had to come into existence simultaneously. And this is exactly what we seem to see in the Big Bang. Contentions that the universe emerged from a quantum fluctuation from a previous state seem ad hoc, forced, and highly speculative. There is no explanation given for any previous universe, or for why it should have given birth to our universe. But the Jewish, Christian, and Muslim doctrine of creation from nothing does nicely explain the ontological origin of the universe, and also its physical origin.

The doctrine of creation explains not only the origin of the universe, but also its incredibly intricate design. One of the main points of Genesis 1–11 is that it is God who is the source of order. Apart from God there is chaos, and ultimately nothingness. We live in a universe in which the physical laws and constants seem to be perfectly attuned and balanced so as to make the universe hospitable to life. While some scientists would explain this as due to chance, the theological explanation is that it is due to God's design. This can be seen in the Genesis account: on the seventh day, God contemplates the creation he has made as a craftsman contemplates a finished masterpiece.

Finally, the doctrine of creation gives us a hint, at least, as to *why* the universe was created. The universe is an expression of God's goodness. It is good in its own right (not just because it is useful for humans), as the author of Genesis 1 repeatedly asserts. But it is good, ultimately, because it is an expression of its maker, God, who alone is wholly good. There is no warrant in Genesis 1–11 for assuming that the universe is made only for the sake of human beings, though human beings are portrayed as having mastery over the other creatures. Rather, it is the human who is the completion of creation. In humanity alone does God find a creature worthy of being a partner in the stewardship of creation. Furthermore, the Christian Scriptures hold out the promise that all things will be fulfilled in Christ. Those worthy of salvation will find eternal peace with God in the resurrection. But the universe itself will share in the glory of the resurrection, according to Romans 8:21: "[T]he creation itself will be set free from its bondage to decay and will obtain the freedom of the glory of the children of God."

God, then, is the all-sufficing explanation for the origin and destiny of the universe. What the Christian explains by God, the naturalist explains by chance. Chance is the origin of the universe, and random decay is its destiny. But chance is really no explanation at all; it is rather the absence of an explanation. So, I would contend, the Christian explanation is more intellectually complete, and more satisfying, than that of the naturalist.

Now the metaphysical naturalist will challenge the theological explanation as adding nothing. Like the "explanation" that the universe was created by a quantum fluctuation, it too is untestable; it too merely pushes the question one step farther back. If God created the universe, he will ask, what created God? Where did God come from? The philosopher Bertrand Russell asks just this question in his famous essay *Why I Am Not a Christian*.[29]

But this is to conceive God as a being on the same level as other created beings, a contingent being, which might or might not exist (as is true of every thing in the universe). However, the idea of God is the idea of One who exists necessarily, who would exist in every possible universe, who cannot not exist. God, says Aquinas, is the only One whose essence is to exist; all other beings (including angels) receive their existence from another. The kind of God that Russell has in mind is not the God of Jews, Christians, and Muslims, but a demiurge, who himself would need to be created, and so exists contingently. Ultimately the claim that God created the universe amounts to the claim that an infinite or eternal series of contingent universes or contingent causes—universes or causes that might exist or might not, but which do not contain their reason for existence within themselves—is impossible. The question as to why anything exists has to terminate in One who contains his (or her) own cause of existence within God's self.

I do not think that this can be proven by reason alone. If it could, faith would be unnecessary. God could be proven in the same way that a mathematical or scientific conclusion can be proven: by reason alone. This kind of proof is simply unavailable, to either side, in the case of ultimate questions, such as "What is the origin of the universe?" "Why does it exist?" and "Does God exist?"

Of what value, then, is theological reasoning? Simply this: faith requires reason as its complement and assistant. If I have faith, but can offer no reason for my faith, even to myself, I am apt to become confused and doubtful, and am apt to get pushed out of my faith by articulate nonbelievers. Furthermore, I will probably not be able to persuade people of my position. Human beings, after all, are to some extent rational and are influenced by reasons. The faithful, then, ought to be able to present a reasonable account of why they believe. This was realized early on in the Christian tradition, when educated Christians such as Justin Martyr, Origen, Augustine, the Greek fathers, and others provided philosophical and theological defenses for their faith. The fundamental doctrines of the Christian church were only arrived at with the aid of philosophical reasoning. Faith in the larger Christian tradition is not blind faith; it is faith supported by and integrated with reason.

Indeed, in the Roman Catholic tradition at least, faith has the capacity to elevate reason so that the reason can understand divine things as well as natural and human things. Put another way, faith expands the horizon of our vision and understanding so that it includes more than what is merely natural and measurable.

I am convinced that Christians can make a more reasonable case for their faith than naturalists can for theirs—that is why I am writing this book. But such a case is cumulative. It depends on a number of converging arguments, entailing not only arguments from nature but also arguments based on human freedom, the human experience of God, the existence of moral absolutes, and so on. The case presented in this chapter is not therefore a stand-alone argument for God; it is one of a complex of arguments that will have to be evaluated by the reader as a whole. I think that a theistic explanation of the origin of the universe makes more sense than a nontheistic explanation, which claims that the universe, in effect, came about by chance—a quantum fluctuation. Each reader will have to decide this for himself or herself. But in deciding, the reader should consider not just the narrow question of the Big Bang, but whether the existence of a personal God also explains better than naturalism other features of the universe and human life, which will be taken up in later chapters.

■ The Complementarity of Scientific and Theological Explanations

Thus the theological explanation of origins answers questions that the scientific account leaves unanswered, even as the scientific account of origins answers questions that the theological account leaves unanswered. Science cannot explain why we have a universe at all, or why we have one structured so precisely as to allow life, or what the purpose of the universe is. Science, by itself, gives us a universe that has no ultimate purpose: an empty, dead universe that came into being accidentally, and in which life has appeared by accident also. Theology by itself, however, cannot explain the history, development, or intricate design of the created universe. It is mute on the question of how the stars, galaxies, planets, and life emerged. It gives the "why" of creation but not the "how." It seems, therefore, that the theological explanation of origins and the scientific explanation of origins complement one another. Each in itself is incomplete; each provides something the other lacks. Only by joining them do we get a complete and satisfactory explanation of the origins of the universe. It is as if two painters were required for painting a picture: one provides the overall framework and background, the other the detailed figures and action in the foreground.

This claim could not be made if the theological account of origins and the scientific account conflicted with each other. But in fact they do not appear to contradict one another, but appear to be consonant with each other. It would certainly be going too far to say that the Big Bang theory "proves" that the universe was created from nothing. The theory cannot prove that there was no previous condition that led to this universe. But as we noted above, at least we can say that if one were to ask what a creation from nothing would look like to an astronomer, the answer is: like the Big Bang. This again underscores the central idea of this chapter: that the scientific account and the theological account are complementary, not competing. Any conflict is due not to divergences between science and theology, but to absolutist ideologies on either side. On the one hand, some theologians think that the Genesis creation account must be read literally, and therefore the Big Bang cannot be harmonized with it and must be rejected. On the other hand, metaphysical naturalists think that everything can be explained by nature, and/or that only scientific, testable theories are real knowledge, and therefore theology has nothing to say. Any conflict, then, is between competing ideologies, not between science and theology.

We will find this same pattern in the following chapters, which deals with the subject of the evolution of life and humanity.

Evolution
The Journey into God

■ The strongest arguments for naturalism have come from evolutionary biologists, such as Richard Dawkins and E. O. Wilson (see chapter 1). We have seen also (chapter 2) that Darwin's work undermined traditional Christian beliefs, e.g., that humans were created directly by God and differed essentially from animals. In the century and a half since Darwin, his argument has been extended to cover virtually every feature of human life. Thus some contemporary evolutionists declare that every human trait, from love and jealousy to morality and religion, can be explained by the blind, material action of evolution. For this reason many Christians have rejected evolution altogether, and embraced so-called "creation science."

I will maintain here, however, that evolution does not have to be understood naturalistically, that is, as a purely natural and unguided process. Instead, I maintain that evolution has a direction, and is ultimately guided by God. But God does not determine every detail of evolution. Rather, God allows an element of freedom (which may appear to us as chance) in that process. Thus both nature and God, or, more exactly, God's Spirit, are causal elements in the process of evolution. Evolution can therefore be seen by Christians as a journey, whose end is the free uniting of persons and creation with God in the resurrection.

Any adequate discussion of evolution is unavoidably technical, and so some of the following paragraphs may be difficult for some readers. But it is not necessary to grasp all the details of the mechanism of evolution to understand the overall argument.

While the fact of evolution is almost universally accepted by biologists, the *mechanism* of evolution—what causes evolution to occur—is disputed. Therefore there is no one theory of evolution. And this suggests that we do not have, presently, a complete theory of evolution, but an incomplete theory, which may be revised in the future, especially as to its mechanisms. Nevertheless, I will argue that, in the broadest terms, evolution means the development of complex systems from simple systems over time. The most important component of this change is that of increasing information: complex systems embody more information than do simple systems. Naturalistic theories of evolution claim that this increasing information is the result of the action of random mutations coupled with natural selection. Theistic theories of evolution, which I will defend here, claim that chance, physical laws, and natural selection, by themselves, are inadequate to account for the emergence of highly complex living systems. While these factors play a large part in the mechanism of evolution, an important factor is also the action of God, whose role is the input of information, which acts to guide the evolutionary process into one path rather than others, and hence into fruitful channels.

Understood in this way, there need be no conflict between evolutionary theory and theism. The real conflict is rather between a purely naturalistic interpretation of evolution and theism.

■ What Is Evolution?

Charles Darwin in his *Origin of Species* (1859) defined evolution as "descent with modification." Living species, he thought, had developed from a few simple ancestral species over vast stretches of time, by a variety of mechanisms, the main one being natural selection. Let us review Darwin's arguments for his theory.

First, he noticed that in any species of plants or animals there was considerable variation. Plant and animal breeders used this variation for their own purposes. If breeders wanted to develop fast horses, they simply bred the fastest horses in any generation, and continued to do this over many generations, until they obtained fast horses. So also with the various desired characteristics of dogs, cattle, pigs, and so on. Any trait could be maximized, within limits, by selective breeding. Darwin called this artificial selection.

Next, Darwin observed that every species of plant or animal leaves vastly more progeny than can possibly survive in the environment. The importance of this was recognized by Thomas Malthus's *An Essay on Population.* Malthus noted that food production increases arithmetically (perhaps a few percent a year) whereas population increases geometrically, or as we would say now, exponentially. The result, according to Malthus, is that the human population will always outstrip its resources, and poverty and famine will be inevitable. The only way to avoid this is to limit births. Darwin realized that this applied also to plant and animal species: "There is no exception to the rule that every organic being naturally increases at so high a rate that, if not destroyed, the earth would soon be covered by the progeny of a single pair."[1] He calculated that even the progeny of a pair of elephants (one of the slowest breeders) would amount to nineteen million in 750 years.

Therefore there exists what Darwin called a "struggle for existence." The fiercest struggle is between plants or animals of the same species. Plants typically produce an abundance of seeds, of which only a tiny percentage reach maturity. Animals, such as fish and frogs, may lay hundreds or thousands of eggs, but only a few reach maturity. The rest are edged out in the struggle for existence. The struggle may also be with predators, or with stringent environmental conditions. A plant on the edge of the desert may have difficulty surviving due to lack of water, rather than due to pests and predators.

The result is that those individuals that are best adapted to the environment will be more likely to survive and leave progeny, while those individuals that are less well adapted, or less "fit," will not survive and leave progeny. This Darwin called "natural selection." This process, he thought, continued over hundreds or thousands of generations, will result in different varieties within one species gradually diverging until they become separate species. Consider the origin of the giraffe. Among horses grazing on the African savanna millions of years ago, some would have longer necks and some shorter. Those horses with longer necks would be able to reach more branches on acacia trees than those with shorter necks. Because there are always more individuals born than the environment can carry, there was a competition for food. Those that could not reach the high branches would perish sooner and leave few progeny; those that could reach the high branches would flourish and leave many progeny. Repeated over thousands of generations, this process resulted in horses with very long necks: giraffes. Note that it is not the case that individual horses stretched to reach the branches, that their necks therefore got longer, and they passed these traits onto their progeny. Acquired characteristics are not inherited. Rather, those horses born with long necks will leave more progeny than those born with short necks. The whole process is automatic and mechanical, yet results in the emergence of a new species, the giraffe.

Darwin's theory is very simple. That is part of its appeal. (Critics would say it is too simple; we will consider this later.)

Modern definitions of evolution vary. According to the National Academy of Sciences, Darwin's definition, "descent with modification," is still a good brief definition of evolution.[2] But it is vague: What is meant by modification? Is it any kind of change? In any direction? What if a species regressed and became a parasite? Would that be evolution, or devolution?

Contemporary evolutionists would probably also want to add that evolution entails the emergence of more complex forms from simpler forms. Life develops from single-celled organisms without nuclei (prokaryotes), to single-celled organisms with nuclei (eukaryotes), and to multicelled organisms, to creatures that can fly, swim, burrow, and run, to intelligent creatures like primates and human beings. Each of these transitions involves an increase in the order of complexity. But what, in this context, is complexity?

This is a term that has proved hard to define. A complex system is one in which many diverse parts and kinds of parts are interrelated into a functioning whole. One way to describe complexity linguistically is that it is a measure of the number of significant interconnections between the parts of a system. Another way to measure the complexity of a system is by trying to approximate the amount of information necessary to describe that system.[3] This, in turn, can be related to computer programming: how long a computer program would it take to adequately describe a given system? Consider a simple example of a complex object: a watch. A watch is complex because it has many differentiated parts, which have to be related to one another in a very specific way. If the shape of even one of the parts is changed, or if the relation of the parts to one another is rearranged, the watch won't function. If we melt the watch down, we will end up with an undifferentiated mass of molten metal and glass, which is relatively simple compared to the watch. It can be described simply in language (for example: a mixture of 51 percent iron, 5 percent chromium, 5 percent silicon dioxide [glass], and so on); it could also be described simply in a computer code. But to describe the watch itself would take a much longer description; each part would have to be exactly specified, and the exact assembly of the whole also specified. Now a sharp critic might say: "Okay, but the specific arrangement of the molecules in the melted-down watch is just as improbable as the arrangement of the molecules in the working watch." That is true. But the point is, one could rearrange the molecules on the melted-down watch in an infinite number of ways, without changing any of the characteristics of the melted mass. But one cannot change the arrangements of the molecules in the working watch in very many ways without destroying the function of the watch. The working watch exhibits what William Dembski calls "specified complexity": complexity, but also a specific arrangement of complex parts, so that the watch works.[4]

Another example of this same principle is language. One can arrange the thirty-five letters and punctuation marks of the English language in many ways, possibly an infinite number of ways. But only a few of these ways will carry a meaningful message. Some arrangements will be ordered but noncomplex, and will carry no message, such as "abcab-cabcabc." Others will be complex, but carry no recognizable pattern, such as "qetuyhedvg.retsd;trds." A few will be complex, but also carry a recognizable pattern, or a message, such as this very sentence. Such sentences are examples of specified complexity.

Now, living systems are vastly more complex than watches. Richard Dawkins notes that each nucleus of an animal or plant cell contains more information than all thirty volumes of the *Encyclopedia Britannica*.[5] The sequence of the nucleotides in the DNA of a cell nucleus must be very precise (at least in some parts of the DNA) in order to code for the formation of proteins necessary for the life of the cell. If this sequence is changed, for example in a mutation, the characteristics of the cell may change, sometimes drastically. Thus in a cell, and in its DNA, we have specified complexity.

Probably the most complex single thing in the universe is the human brain. It has been compared to a forest covering the western half of the United States, of 10,000 trees per acre, each with 10,000 connections with other trees. The human brain is much smaller than a star, but is a more complex system, because there are far more significant interconnections between its parts than in a star. There are many more atoms in a star, but fewer types of atoms, and each has relatively few significant interconnections. We can mix up the parts of a star without destroying the star, but not those of the brain without destroying it. Since the number of significant connections is greater than those in a star, it would take many times more lines of computer code to specify the significant connections between the molecules of a brain, or a person, than between those of a star. In a star, we do not have a high degree of specified complexity, whereas in a person, we do. "Complexity" in evolution, therefore, is shorthand for "specified complexity," which is essential for living things. And a high degree of specified complexity corresponds with a high degree of information, because it would take a large amount of information to specify it.

In addition to describing evolution as "descent with modification," then, we will include in our definition the development of complex systems from simpler systems over time. Note that this definition avoids the whole problem of species. Species, in modern biology, are not seen as natural types, but as particular adaptations to environments, adaptations that are subject to change. A change in the shape of the bill in a group of finches may constitute a new species. But it would not necessarily correspond to an increase in genetic information.

■ Neo-Darwinism

There were many problems that Darwin was unable to solve. He did not know the source of variations in a population, or how variations are passed on. One of the arguments against his theory at the time was that any advantageous variation would soon be swamped by interbreeding in a large population, and fail to leave any mark on the population as a whole—like mixing in a tiny bit of black paint into a large pot of white paint. The full answer to this problem for Darwin's theory was not possible until the emergence of the modern science of genetics, which also explained the source of variations. We now know that heritable variations are passed on by genes, and genes are discrete units, which retain their distinctiveness in the genome even if they are not expressed. Thus, for example, the gene for blue eyes is recessive. If brown-eyed Anne marries brown-eyed Steve, and they both carry a recessive gene for blue eyes, some of their children may have blue eyes. Their blue-eyed children will not have eyes that are a blend of brown and blue, because the gene for blue eyes, though recessive, does not blend with the brown-eyed gene and become diluted.

Genetics also explained the source of variation. Genes are composed of long sequences of DNA. These sequences can be changed by several means. One is that in replication, copying errors are made, so that the "child" DNA does not have exactly the same sequence as the "parent" DNA; there may be a difference in only one nucleotide, but that may mean the gene codes for a different protein. Or the DNA sequence can be changed as a result of chemical pollution, or from radiation, either from space (cosmic rays) or from other sources. Any change in the DNA sequence in the germ cells (sperm or eggs), unlike changes in somatic cells, may be passed onto offspring.

The combination of modern genetics with Darwin's original theory of evolution is termed "neo-Darwinism." The modern "synthetic" theory incorporates a number of other changes from Darwin, for example "genetic drift"—the idea that genes in a population may change randomly and in no particular direction even if they are not subject to natural selection. But insofar as the modern theory still places the principal agent of the origin of species in random mutation and natural selection, it is still considered neo-Darwinism. We will see below, however, that there is considerable disagreement as to how much of a role natural selection really plays in evolution, and what some of the other mechanisms of evolutionary change might be.

■ The Mechanisms of Evolution

It is important to distinguish between the simple fact of evolution and the mechanisms of evolution: how evolution actually happens. There is virtually no disagreement among biologists that evolution has occurred. But there is considerable disagreement on the mechanisms of evolution. Some, like Richard Dawkins, argue that random mutation and natural selection are the only mechanisms by which adaptive complexity can be achieved. Others, like Stephen Gould, place more emphasis on chance, but also on selection operating at many levels (gene, organisms, etc.). Steven Rose argues that organisms must be considered first of all as wholes, which affect their environment as well as being affected by it. To some extent, they create their own lifelines and even their own evolutionary trajectories. Stuart Kauffman argues that natural selection must be supplemented by processes of self-organization. Whereas most evolutionary theorists emphasize the role of chance and contingency in the path of evolution, Michael Denton argues that the biosphere, as well as the universe, is structured so that the emergence of carbon-based life is all but inevitable, and that the evolutionary path itself is largely preordained. And these are only some of the contending positions. By and large, orthodox neo-Darwinism has won the university faculties in North America, but there are dissenters scattered all across America and the globe. I shall survey briefly here the major positions, and conclude that actually there is not one theory of evolution, but many. Consequently, evolution is an incomplete theory. This means that it is risky to base far-reaching theological and philosophical conclusions on it. In one hundred years, we may have a very different theory. Finally, I will outline a theistic account of evolution, and argue that it explains the facts better than a naturalistic account.

Let's begin with what is usually called orthodox neo-Darwinism. Its central idea is that evolution is driven by random mutations and natural selection. Mutations are changes in the sequence of the nucleotides in the DNA. A change in even one of these can change a gene and the protein that the gene codes for. Mutations might be caused by background radiation, by cosmic rays, or by errors in copying the DNA when the germ cell divides. Only mutations in the germ cells (sperm, eggs) used in reproduction have an effect on offspring; mutations in body cells have no such effect. In the germ cells, any mutation will be passed on to offspring. Mutations are said to be random, which means random with respect to the good of the organism. By far the majority of mutations are harmful. Some are lethal, like the mutation that causes frogs to have extra legs. But it is essential to the neo-Darwinist account that some mutations are beneficial, that is, they help the organism adapt to the environment, survive, and leave more progeny. Thus a fly may have a mutation that confers a partial resistance to DDT. This would only help the fly in an environment that contains DDT, but there such a mutation could be crucial to surviving at all, and therefore to leaving progeny.

Now it is part of the neo-Darwinist paradigm that mutations only change organisms by tiny degrees, never by saltation (i.e., jumps). Natural history does not go from a fish to an amphibian or from a reptile to a bird in one jump, but by degrees. And—this is important—each degree of change, if it is to survive, must confer some advantage to the organism in the struggle for existence. So, to go from scales to feathers, there must be a whole series of tiny changes, each of which confers a survival advantage on the organism. Darwin himself details one example of evolution, namely, the case of the eye. Organisms do not go from being eyeless to having fully functional eyes in one step. Rather, they first develop simple light-sensitive spots. Then, in succeeding generations, each spot deepens into a pit and the rims of the pit close, forming a simple pinhole camera eye. Later stages add lenses that can focus, more advanced retinas, and so on. A fully functional vertebrate eye emerges only after thousands of generations. Various groups of animals can be found that have all different stages of eyes, from rudimentary light spots, to pinhole cameras, to lenses, to advanced eyes.

The most compelling (and triumphalist) account of neo-Darwinism is provided by Richard Dawkins, whose view of evolution has been laid out in a series of popular books: *The Selfish Gene*, *The Blind Watchmaker*, *Climbing Mount Improbable*, *River out of Eden*, and others. Dawkins's version of evolution is based squarely on the agency of natural selection:

> Natural selection, the blind, unconscious, automatic process which Darwin discovered, and which we now know is the explanation for the existence and apparently purposeful form of life, has no purpose in mind. It has no mind and no mind's eye. It does not plan for the future. It has no vision, no foresight, no sight at all. If it can be said to play the role of "watchmaker" in nature, it is the *blind* watchmaker.[6]

Selection for Dawkins is at the level of the gene, and his vision is resolutely reductionistic: "a central truth about life on earth . . . is that living organisms exists for the benefit of DNA rather than the other way around."[7] Human beings are no more than machines created by genes:

"We are survival machines, robot vehicles blindly programmed to pre-serve the selfish molecules known as genes."[8] Evolution itself is due to the accumulation of millions of tiny changes (themselves due to random mutation), which are sorted and directed toward adaptive complexity by the blind activity of natural selection.

On the other side of the Atlantic is Dawkins's rival, the American pale-ontologist Stephen Jay Gould (now deceased). Gould was famous as one of the developers of the theory of punctuated equilibrium. This view of evolution emphasizes that the fossil record reveals not so much gradual development, as would be expected from Darwin's and Dawkins's view of evolution, as long periods of stasis (equilibrium), punctuated by sudden extinctions, or the apparently sudden emergence (in geologic time) of new species.[9] For example, around the beginning of the Cambrian epoch (about 600 million years ago), all the basic animal phyla, each exhibiting a different body plan, suddenly appear in the fossil record. Gould calls this the "Cambrian explosion." It was followed by massive extinctions of many of the phyla, and then later diversification within the remaining phyla. The result is that the plan of animal history is not a simple tree diagram, in which a few simple forms gradually diversify into many diverse and complex species. Darwin himself had sketched out such a plan in his *Origin of Species* as that which would be expected on the basis of his theory. In standard evolutionary textbooks, this has become known as the "cone of increasing diversity." From simple origins, diversity increases over time. But the Cambrian explosion belies this. Instead, we find extensive diversity right at the beginning, followed by decimations as phyla are extinguished, and then subsequent diversification within the remaining phyla. Gould provides the example of the echinoderms. There are now five groups of this marine phylum: sea starts (starfish), brittle stars, sea urchins, sea lilies, and sea cucumbers. All show fivefold radial symmetry. But fossils from the time of the Cambrian explosion show twenty to thirty groups of echinoderms, some with body plans very different from any modern representative.[10]

Now, it is possible to explain this on the standard neo-Darwinist model: one merely has to assume that there is a long history of precambrian life which diversified into twenty or thirty or more different phyla, but left no fossil traces. But still, this pattern is not what one would expect if the orthodox neo-Darwininan model, championed by Dawkins and others, were correct. One would expect the tree diagram, the "cone of increasing diversity."

Gould has also persistently argued against Dawkins on other points as well. First, he contends that natural selection acts at many different levels of "Darwinian individuality (from genes to organisms to demes to species to clades)," and not just at one level, the gene.[11] Second, with his colleague Richard Lewontin and the British biologist Steven Rose, he insists that organisms shape their own environments, and are consequently not simply passively shaped by natural selection. Natural selection, Gould writes, is "a necessary but by no means sufficient, principle for explaining the full history of life."[12] Third, much more than Dawkins he opts for the role of chance in evolution. His book *Wonderful Life* argues that massive extinctions have played a major role in shaping the direction of evolution, and that the survival of species through these events is truly random, not the result of natural selection. "The history of life is a story of massive removal followed by differentiation within a few surviving stocks, not the conventional tale of steadily increasing excellence, complexity, and diversity."[13] Thus, if the tape of evolution were played back again, it would in all probability reveal a vastly different scenario. Gould is hostile to any notion of progress in evolution. Advances toward complexity, he says, can be balanced by regression from complexity, as in the case of parasites; progress is in fact a statistical illusion, fostered by humanity's anthropocentric hopes.

A number of other biologists also have challenged the centrality of natural selection. Prominent among these is Stuart Kauffman, a member of the Santa Fe Institute. Kauffman's core idea is that, besides random mutation and natural selection, there is another source of order in biological processes: self-organization. Complex systems, from computer-modeled networks to biological systems, exhibit a surprising degree of spontaneously generated, emergent order (simple examples of this, in physics, are the hexagonal structure of snowflakes or the cone shape of sandpiles). Kauffman holds that self-organizing processes are the origin of life itself, and the origin of much of the order in living systems generally. He explains:

> I propose that much of the order in organisms may not be the result of selection at all, but of the spontaneous order of self-organized systems. . . . The order of organisms is natural, not merely the unexpected triumph of natural selection. For example, I shall later give strong grounds to think that the homeostatic stability of cells, the number of cell types in an organism compared with the number of its genes, and other features are not chance results of Darwinian selection but part of the order for free afforded by the self-organization in genomic regulatory networks. If this idea is true, then we must rethink evolutionary theory, for the sources of order in the biosphere will now include both selection *and* self-organization.[14]

Kauffman's work has influenced a number of biologists, such as Brian Goodwin, Mae Won Ho, and Steven Rose, all connected with the Open University in Britain.[15] The focus of their biology is on the centrality of the whole organism and its processes of development rather than the gene. David Depew and Bruce Weber have this to say about the effects of this new thinking in developmental biology, which is challenging the hegemony of Darwinian orthodoxy.

> By contrast to Darwinism's stress on random variation and selection by external, environmental forces, an older and surprisingly persistent developmentalist tradition has consistently maintained that the evolution of kinds (phylogeny) should be viewed as an inner driven process like the unfolding of an embryo (ontogeny). . . . The developmentalist tradition displaced Darwinism in the later nineteenth century, but was in turn marginalized by it for most of the twentieth century. . . . [L]atter day developmentalists [such as Brian Goodwin, Mae Won Ho, et al.] have been asserting of late that natural selection, and so Darwinism, will not survive the transition from the "sciences of simplicity" to the "sciences of complexity."[16]

Stephen Gould has been sympathetic to these developments. He writes: "The constraints of inherited form and developmental pathways may channel . . . changes so that even though selection induces motion down a permitted path, the channel itself represents the primary determinant of evolutionary direction."[17]

British biologist Steven Rose also emphasizes the centrality of organisms and their "lifelines," and is an acerbic critic of Dawkins and of genetic reductionism in general. Rose, a Marxist, argues passionately for "making biology whole again." The kernel of this program is the idea that within the world different levels of organization emerge, which have to be explained by different methods.

> So at each level different organizing relations appear, and different types of description and explanation are required. Hence each level appears as a holon—integrating levels below it, but merely a subset of the levels above. In this sense, levels are fundamentally irreducible; ecology cannot be reduced to genetics, nor biochemistry to chemistry.[18]

Living things are conditioned by their components, but also by their environments and contexts, and by their unique histories. Furthermore, organisms affect their environment even as the environment affects them. The future of evolution, then, is radically unpredictable. Humans are the more or less accidental product of evolutionary forces, but are nevertheless free to some extent to construct their own future.[19]

Robert Wesson argues that conventional neo-Darwinism is too simple and too mechanistic an explanation for evolution—it is the parallel of Newtonian physics in biology. Recent scientific advances, especially in chaos theory and theories of self-organization, are necessary to get a better understanding of evolution. Evolution is the story of the development of complex systems, exactly the kind of systems described by chaos and complexity theory. Wesson sees the emergence of stable genomes and stable biological patterns as examples of what are called in complexity theory "attractors." He writes:

> Any self-reinforcing pattern, or a self-ordering part within the self-ordering whole, such as the brain . . . may be considered an attractor. Any taxonomic group, as far as it represents a set of organisms that belong together, represents an attractor at its level.[20]

Complex dynamic systems, stabilized by the presence of attractors, resist change (this is the origin of stasis in species), but when they do change, they change not gradually, but suddenly, as they shift to another stable pattern. This might help explain the origin of speciation. Wesson concludes that "Evolution is the result of at least four major factors—environment, selection, random-chaotic development, and inner direction—and one might no more expect to find any law to govern it than to find a law of the mind."[21] Wesson does think that evolution progresses in the direction of complexity and biodiversity, and that biological evolution gives way to cultural evolution in human societies.

Now, admittedly the ideas of Kauffman, Goodwin, Ho, Rose, Lewontin, Depew and Weber, and Wesson are somewhat outside the mainstream of contemporary neo-Darwinian orthodoxy, which typically explains evolution by random mutation, natural selection, genetic drift, and chance factors (for example, a sudden change in the environment). Indeed, most biologists, concerned as they are with microproblems (see the articles in any issue of the journal *Evolution*), have probably not read these authors, most of whom are, however, practicing biologists. But even a slight knowledge of the history of science will teach us that it is folly to ignore minority ideas, for sometimes they prove to be the key to scientific discovery. (A striking example is the theory of continental drift, which was ridiculed for decades, yet is now one of the foundational theories in modern geology.)

The point of this brief survey is to demonstrate that, while there is consensus on the fact of evolution, there is disagreement on the mechanism of evolution—what drives it—and on whether, or in what sense, it might be progressive. Almost no practicing biologist would argue that evolution has a preordained target (for example, humanity). But some, like Dawkins or Wilson, argue for a kind of progress from simplicity to complexity, while others, like Gould, argue against any progress at all. The resurgence of developmental thought based on self-organization and nonlinear dynamics may forecast a major change in evolutionary theory.

From this, I draw a provisional conclusion: we do not now have a finished theory of evolution. Perhaps, as Depew and Weber suggest in their book *Darwinism Evolving*, it is still evolving. But this means that any *theology* of evolution must necessarily be tentative, for the shape of evolutionary theory one hundred years from now may be quite different than it is now.

There is another reason to suspect that this might be the case. The history of science over the last few centuries shows clearly that scientific theories are provisional and tentative, and may be subject to drastic reformulation even after having commanded a consensus for decades, or even for centuries (as in the case of Newton's theory). A well-known example of this is the fate of the theory concerning the ether. The existence of the ether, as a necessary medium to transmit light waves, was unquestioned by physicists for most of the nineteenth century.[22] Indeed, James Clerk Maxwell, one of the giants in the history of physics, wrote the article on the ether for the ninth edition of the *Encyclopaedia Britannica* (1878), in which he calculates the coefficient of rigidity and density of the ether to two decimal places. He concludes the article with these words:

> Whatever difficulties we may have of forming a consistent idea of the constitution of the ether, there can be no doubt that the interplanetary and interstellar spaces are not empty, but are occupied by a material substance of body which is certainly the largest, and probably the most uniform body of which we have any knowledge.[23]

This was written about 124 years ago. And yet today, the theory of the ether has only historical interest. It has gone the way of the phlogiston theory and the geocentric universe.

Thus it is risky to speculate on what the theory of evolution might look like 124 years from now, especially with potentially revolutionary developments on the horizon.

■ Directed Evolution

All the theories considered above are examples of *undirected* evolution, that is, an evolutionary process that is not directed by any form of intelligence. Recently, New Zealand microbiologist Michael Denton has proposed a theory of directed evolution, that is, evolution directed by intelligent agency. It is this theory that I wish to consider here.

Denton first argues that the biosphere of the earth is uniquely suited for life, and even evinces evidence of having been designed for life. He marshals an impressive amount of data to support this claim, but I can only give one example here: water. Water has a unique set of physical properties that allow it to support life. It contracts as it cools, down to 4° Centigrade, then, as it cools more, it expands. Furthermore, ice expands as it is formed. If this were not so, much of the water on the earth would be locked up in vast beds of ice under the lakes and oceans, which would freeze from the bottom up. In such an environment, it would be difficult for life to develop or thrive. The latent heat of water (i.e., the amount of heat that it takes to freeze or evaporate a given amount of water) is one of the highest known. This allows bodies of water to stabilize the climate, and also allows warm-blooded animals to rid themselves of their excess body heat by sweating. Water has high thermal conductivity, which allows for equal distribution of heat throughout the bodies of animals. And so on.[24]

This kind of "fine tuning" of the environment so that life might exist leads Denton to conclude that the universe is designed so that life, and specifically intelligent life, might emerge by a process of directed evolution. This thesis is an extension of the anthropic principle, encountered in the previous chapter, from the area of cosmology to the physics and chemistry of life on earth. Of course, a skeptic might reply that the earth is peculiarly fit for the existence of life because if it were not, we wouldn't be here to write about it. All that Denton's argument shows is that one planet, out of trillions possible, had the requisite conditions for the emergence of life. But Denton's argument is deeper than this. He contends that it is the universe, not just the earth, that is uniquely suited for life. The properties of water that make it suitable for life are not unique to earth; they belong to the physics of the universe.

Denton in addition provides specifically biological arguments for directed evolution. The action of most genes, he writes, is pleiotropic; that is, single genes typically affect many different structures or processes in an organism. It is not usually the case that one gene affects only one structure, so that, for example, there is a one set of genes governing the development of the heart, another for the brain, another for the liver, and so on. Rather, many genes in different groups affect the development of any one organ and, usually, each gene affects many characteristics in different organs. Examples like a single gene controlling eye color are the exception, not the rule. One of the best illustrations of pleiotropy is the genetics and development of the nematode. Nematodes are tiny worms existing in garden soil, the human body, and other places. They are simple organisms and have been studied in great detail. Biologists now understand the developmental history of each of the 959 cells in the nematode larva; that is, they can trace the development of each of these cells back to the genes that govern that development. One might expect that each organ in the nematode would be specified by a particular set of genes that stipulated the proteins and developmental pathways for that organ and no other, just as one work unit in an automobile factory might produce engine blocks, another carburetors, another generators, and so on. But the structure of the nematode is not nearly so simple. Each organ is influenced by many different, unrelated genes, each of which influences many different organs. It is as if, in an automobile factory, each worker worked on many different kinds of parts, such as engines, carburetors, generators, headlights, etc., and each such assembly was not the product of one discrete work unit but of many workers from across many different areas. Denton writes: "No structure or process or organ is genetically isolated; the same genes are involved to a greater or lesser degree in the specifications of all organs, structures, and processes."[25]

This means that the nematode is structured not on the basis of modules, like a machine, but holistically, like an organism (which it is).

> There was not the slightest evidence for genetic modules—that is to say the existence of groups of genes involved in the specification of particular parts of the organism, say, the nervous system or the pharynx. No matter which part or organ of the nematode one examined, its development was intimately interconnected with the development of practically all the other organs and parts of organs.[26]

The equation of organisms with machines, favored especially by Richard Dawkins and his ilk, turns out to be false.

Now this is especially significant for understanding the mechanism of evolution. If organisms are structured so that, typically, each gene affects multiple characteristics, it is much more difficult for organisms to evolve by changing one gene, or one nucleotide in one gene, at a time. For changing one gene will affect many different systems, each of which is already well adapted to the whole organism and to its environment. Even if the change is beneficial for one system, it will almost certainly be less than beneficial for other systems, and so result in poorer adaptation or fitness overall. To significantly change an organism, it would be necessary to change many genes simultaneously in very specific ways. But this is precisely the kind of saltation that orthodox neo-Darwinism rules out, because the odds against changing many genes simultaneously, each in a highly specific way, by an undirected, random process, so as to achieve a better fit to the environment, are simply astronomical. It would be like changing a jigsaw puzzle; any change to one part entails changes to many adjacent parts, and these changes ramify through the whole puzzle, so that, in effect, many or most pieces would have to be changed simultaneously to retain an integrated puzzle.

Interestingly enough, this was precisely the argument that Georges Cuvier, one of the leading anatomists in early-nineteenth-century Europe, used against Darwin's theory: "All the organs of one and the same animal form a single system of which all the parts hold together, act and react upon each other; and there can be no modifications in any one of them that will not bring about analogous modification in them all."[27]

Stuart Kauffman, whose work we have discussed above, argues in a similar vein:

> I believe this to be a genuinely fundamental constraint facing adaptive evolution. As systems with many parts increase both the number of those parts and the richness of interactions among the parts, it is typical that the number of conflicting design constraints among the parts increases rapidly. Those conflicting constraints imply that optimization can attain only poorer compromises. No matter how strong selection might be, adaptive processes cannot climb higher peaks. . . . [C]onflicting constraints are a very general limit in adaptive evolution.[28]

Another genetic phenomenon that implies the same conclusion is that of genetic redundancy—that is, a structure or process is controlled by two separate genetic programs, either one of which would be sufficient to produce the structure or process by itself. The best example of this, according to Denton, is the vulva in the female nematode:

> the organ is generated by means of two different developmental mechanisms, either of which is sufficient by itself to generate a perfect vulva. . . . Now this phenomenon poses an additional challenge to the idea that organisms can be radically transformed as a result of a succession of small independent changes, as Darwinian theory supposes. For it means that if an advantageous change is to occur, in an organ system such as the nematode vulva, which is specified in two completely different ways, then this will of necessity require simultaneous changes in both blueprints.[29]

Now, this does not mean that flies cannot adapt to DDT or bacteria to penicillin, or finch populations on the Galápagos cannot develop different shapes of beaks, by means of natural selection. Such changes do not require radical alterations in the organization of the genome of the organism. They are examples of microevolution, not macroevolution. Neo-Darwinism argues that macroevolution is just an extended trajectory of microevolution, which eventually results in greatly different species, different body plans, and so on. But if what Cuvier, Kauffman, and Denton hold is true, then a radical change from one kind of species to another could not be achieved by piecemeal and tiny genetic changes. What would be required would be a change in many genes at the same time, a holistic change. This is also shown by patterns of artificial selection. Breeders have succeeded in creating many different varieties of dogs, pigs, pigeons, horses, etc. But they have not been able to go beyond the species level; these species, at least, exhibit a holistic kind of stability that cannot be changed by small incremental changes.

We can see a similar pattern in human inventions. Once the general conception of a bicycle, automobile, computer, or airplane has been arrived at, they can be gradually perfected by small, incremental improvements. But the original idea of, say, a bicycle or an airplane is not arrived at by small, incremental changes, but as a leap to a new concept. That is why it is called an "invention" rather than an "improvement."

Now this also fits the fossil evidence, which is one in which stasis predominates; that is, new species appear suddenly in the fossil record, exist unchanged for millions of years, and then disappear. Niles Eldredge, a paleontologist, writes: "Gould and I claimed that stasis—nonchange—is the dominant theme in the fossil record. It is characteristic of most species that have ever lived. Adaptive change is relatively rare."[30] To be sure, gradual change occurs in the fossil record. But, as Eldredge observes,

> the point is that such gradual change . . . never seems to *get* anywhere. . . . [M]ost gradual change . . . in the fossil record seems to be more a to-ing and fro-ing—a sort of oscillation within a spectrum of possible states. Typically, a lineage will get larger for a while, then start to get smaller.[31]

This is due to what is known as "stabilizing selection." If a species is well adapted to its environment, any change may make it less fit, hence those individuals of a population who vary greatly from the mean, like frogs with six legs, will be weeded out by natural selection. This notion is commonplace in biology, but it does raise a problem. Natural selection (working on random mutations) is said both to shape new species and to stabilize existing ones. In other words, it both fosters change and resists change. The usual explanation for this is that speciation (the development of new species) takes place in reproductively and geographically isolated populations, whereas stabilizing selection is dominant in large populations in relatively unchanging environments. But the fact of stasis, and the idea of stabilizing selection, makes even more sense if coupled with Denton's contention that radical species change requires multiple simultaneous genetic changes, events that would be quite rare.

■ Directed Evolution and Chance

Directed evolution implies that some sort of intelligence directs the general course of evolution. Denton notes that this could be a god, or a Platonic "world soul." But even if one believes in directed evolution, and Denton makes a good case for it, that does not rule out the presence of chance in the evolutionary process. By "chance" here I mean the intersection of two unrelated causal chains. A boulder, which has rested on a hill since the last glaciation, suddenly rolls down the hill because of erosion, and happens to collide with a car on the road below, killing the occupant (this actually happened in St. Paul some years ago). There appears to be no relation between the two causal chains, yet they intersect, and tragedy results. Now this kind of chance is a pervasive feature of our lives and of nature. If we say that God controls every event, then we have to say that God caused the boulder to roll down at the precise moment when it would hit the car. God is therefore directly responsible for evil (that is, a lack of a good that should be present), a conclusion that most Christian theologians wish to avoid. It seems more reasonable to say that God allows a certain amount of chance or freedom in nature as a way of letting nature become itself. Consider the example of snowflakes. The fundamental hexagonal structure of the flake is given by the molecular structure of water, and hence by the laws of physics and chemistry. But snowflakes vary enormously in detail, as anyone knows who has ever looked at photos of snowflakes. The exact detail of the crystallization of the water molecules on each arm of the flake is a matter of chance—how many water molecules happen to be in the vicinity of the flake, the path of its fall, and so on. Does God specify or guide the development of every snowflake? I cannot believe that he does. Rather, God leaves an element of freedom (which appears to us as chance) in the workings of nature. In this way very great variety and beauty develop within a framework of given physical laws. The colors of the autumn foliage, the shape of clouds, the play of light in a sunset, each can be understood as the result of chance working within a framework of physical law.

But of course the presence of chance also means that evil as well as good can result. Most chance mutations are harmful. Genetic defects in children or adults can result in terrible suffering. Earthquakes and other natural disasters are also the product of chance in the sense defined above. To eliminate such evils, God would either have to eliminate sentient and conscious beings, and hence eliminate much good, or control every detail of the creation and eliminate freedom, both of conscious creatures and of the creation itself, again entailing a loss of much good.

I will have more to say about the evil that seems to be an intrinsic part of the evolutionary process below.

Many of the phenomena of evolution, which do not require large inputs of information, can be explained by the operations of chance mutations and natural selection. The various bill shapes of the finches in the Galápagos do not involve large changes in the finch genome, and can be easily explained by mutation and natural selection, and need not be seen as examples of directed evolution.

■ Natural Selection and the Creation of Information

The standard argument of neo-Darwinian orthodoxy is that the combination of random mutation and natural selection is what creates the buildup of information that is necessary for the emergence of complex forms of life. The question is: Can the passive action of the environment, "acting" through natural selection by a series of mutations and tiny incremental steps, build up the information in complex living systems?

In a recent book, *Finding Darwin's God* (quoted and referenced below in this paragraph), biologist Kenneth Miller gives an example of the creation of new information by the evolution of bacteria. He cites the experiments of Barry Hall, published in 1982 in an article entitled "Evolution on a Petri Dish."[32] Hall was studying how the bacterium *Escherichia coli* metabolized lactose (milk sugar). He deleted from the bacteria being studied the structural gene that codes for the production of the enzyme galactosidase, which allows the bacteria to metabolize lactose. Then he challenged these bacteria to grow on a culture of lactose. At first they could not grow, since they now could not produce the enzyme that would cleave the lactose into sugars, which they could metabolize. But after some days, strains of bacteria emerged that were able to metabolize the lactose. Miller comments:

> How could these cells have reconstructed the information from the missing gene in such a short time, using only the random undirected processes of mutation and natural selection? . . . They made it by tinkering with another gene, in which a simple mutation changed an existing enzyme just enough to make it also capable of cleaving the bond that holds the two parts of lactose together.[33]

But this would not be enough unless this gene, which is normally inactive, was turned on by its own regulatory gene when in the presence of lactose. Hall discovered that this regulatory gene also had mutated so that it switched on the substitute gene when the bacteria was in the presence of lactose. Thus, to evolve a system capable of metabolizing lactose, there had to be *two nearly simultaneous* mutations, not just one. This experiment seems definitive proof that random mutation and natural selection can reproduce deleted information. Evolutionary biologist Douglas Futuyma concludes just this: "One could not wish for a better demonstration of the neoDarwinian principle that mutation and natural selection in concert are the source of complex adaptions."[34]

However, when one goes back and reads the original article, one finds that the situation is not so simple. First, the "substitute" gene that was mutated to produce the galactosidase was already present in the bacteria, and apparently served no function (Hall calls it a "cryptic gene"). The mutation changing it was relatively minor (Miller refers to the change as "tinkering"), and it is not clear that that mutation increased the information available to the bacteria. This substitute "cryptic" gene may in fact have been an example of biological redundancy, like the two separate systems that specify the vulva in the nematode (above). If organisms have built-in redundancy, then there is much less chance that an accidental crippling of an adaptive system, as Hall induced in the *E. coli*, would exterminate these organisms.

In another experiment, Hall found another example of a redundant "cryptic gene" when he challenged a certain strain of bacteria to grow in salicin, which is a nutrient that they could not normally grow on. Again, through two mutations, the bacteria "learned" to grow in salicin by activating the dormant "cryptic" gene and its corresponding regulatory gene. Thus it seems likely that bacteria and possibly other living organisms have a lot of reserve DNA, coding for genes that are normally not used, but that can be switched on through mutations if the bacteria are in an environment in which such genes become necessary for survival.

There is a second point, which is even more interesting. One of Hall's findings was this: the mutations that allowed the *E. coli* to metabolize lactose (after the gene for galactosidase had been excised) were apparently *not random*. How could he tell? Recall that two mutations were necessary: one to the substitute "cryptic" gene so that it would code for galactosidase, another to its corresponding regulatory gene. Hall calculated that the random chance that both of these mutations would occur together in the same bacterium was 1 in 10^{18}. In other words, some 10^9 cell divisions would have to occur for the first mutation to occur (if it were entirely due to chance), and some 10^9 cell divisions would have to occur in these cells before the second mutation occurred. The chance of this is vanishingly small. Hall comments:

> If, as we would expect, the mutation to *ebgA+* is neutral, then, given a steady state population of 10^{10} cells per colony, including ten *ebgA+* cells,

then some 10^5 [100,000] years would be required for 10^9 cell divisions of *ebgA+* cells.[35]

But instead of 100,000 years, bacteria with both mutations appeared in 9 days. Hall concludes as follows:

> We have shown that the selective medium is not itself mutagenic, and that the strain does not carry mutator mutations, since the colonies exhibit no higher frequency of mutations to drug resistance than expected. We can only conclude that under some conditions spontaneous mutations are not independent events—heresy, I am aware.[36]

Now of course this conclusion has been challenged. In an article in *Science* in 1993, Richard Lenski and John Mittler argue that there were insufficient controls in Hall's experiment. It is still possible that one or another undetected factor could have caused the high incidence of mutations. But they acknowledge that a number of experiments have shown that the rate of favorable mutations seems to be much higher under selective than under nonselective conditions. (Selective conditions occur when, for example, bacteria that cannot grow on a certain nutrient [for example, lactose] are put on Petri dishes in which only that nutrient is present.) This is known as the problem of "directed mutations." It seems to imply that under starvation conditions (i.e., selective conditions) bacteria might be able, in some sense, to "choose" or direct their mutations so as to be able to survive. But this violates one of the basic tenets of neo-Darwinism: that mutations are undirected and random, that is, they have no relation to the welfare of the organism. Lenski and Mittler conclude that the case for directed mutations has not been proven, and that there is no reason thus far to change this basic tenet of neo-Darwinism. But they also acknowledge that it remains a possibility, supported by considerable experimental evidence: "Still, the view that certain mutations happen much more often when they are advantageous has been accepted by some observers."[37]

This means that we cannot say, as Miller does (above), that Hall's experiment, or any other, has demonstrated that *random* mutation plus natural selection has the capacity to build up information in organic systems. First, in Hall's experiment, the information was largely there already, in the form of "cryptic genes." Second, the mutations appear to have been nonrandom; such was Hall's own conclusion. If the mutations are nonrandom, we simply cannot say where they came from. The simplest hypothesis is that they are in part somehow directed by the organisms themselves. This is "heresy" (as Hall noted), because neo-Darwinian orthodox theory does not admit that organisms can affect their own genetic structures. It is also possible, if mutations are directed rather than undirected, that they must be directed by some kind of higher intelligence. This would result in the possibility of "directed evolution," which I will consider below.

A strong case against the belief that information is generated by random mutation and natural selection has been made by physicist Lee Spetner, an expert in information theory. In an extensive survey of information theory in biology, Spetner shows that in most cases of adaptation, for example, the adaptation of a bacillus to penicillin in the environment, information is not gained in the genome; it is lost. This is because an antibiotic works by fitting a matching site on the ribosome of a bacterium and so interferes with its ability to make a certain protein, thus impeding the function of the bacterium. The bacterium acquires resistance to the antibiotic by a point mutation, which makes the matching site less specific, so that the antibiotic cannot attach to it. A reduction in specificity corresponds to a decrease in information.[38]

I conclude, then, that the neo-Darwinian claim that information is built up in the genome by random mutation and natural selection has not been demonstrated. Other possibilities remain. Information seems to emerge as a result of a *nonrandom* process, that is, directed mutations, or directed evolution.

■ Directed Evolution and Intelligent Design

Authors in the intelligent design movement, such as William Dembski[39] and Michael Behe,[40] assert that at least some features of the natural world exhibit features of design. Behe, for example, argues that structures such as the cilium of a swimming cell or the flagellum of a bacterium are *irreducibly complex*. Like a mousetrap or a watch, they only work if all the parts are present and arranged in a specific order. Move or remove one part, and the trap or the watch does not work. Behe argues, therefore, that such biological structures could not have been created stepwise, as the Darwinist scenario requires. (Behe's conclusions, however, have been vigorously disputed by most biologists.)

Authors such as Dembski and Behe are usually understood by their critics as holding that God must have reached into the evolutionary process at some point and inserted a designed organism or biological structure. Biologist Kenneth Miller, in a lengthy critique of Behe's position, reads Behe this way. Now, I do not want to argue here either for or against the presence of irreducibly complex biological entities—they may or may not exist. This is a highly technical question—Could such and such a structure possibly have been built up in a stepwise fashion, in which all steps contribute to the fitness of the organism?—that is best left to professional biologists. But if there are irreducibly complex biological structures, they are in any case more easily explained by a process of directed evolution than by direct divine intervention. Directed evolution could also create structures of irreducible complexity, but would do so through a gradual evolutionary process, as a change from a simpler ancestor to a more complex descendant.

An argument frequently voiced against any claim of design in evolution is that while many biological structures do appear as if they were designed, many do not. Stephen Gould gives as an example the panda's thumb. The thumb of the giant panda bear is not one of the five digits of the hand, but an elongation of the radial sesamoid bone of the wrist. Gould characterizes it as a "contraption" not a "contrivance," and his point is that it is a kind of jerry-built structure that no designer, if designing from scratch, would ever construct.[41] But the point is that in any process of evolution, undirected or directed, does not create organisms *de novo*, but must perforce work with what is already there. If, across many millions of years, a population of amphibians evolves from a population of fish, the evolved population will necessarily share most of the features of the parent population, including most of its genetics and biochemistry. Evolution does not create from nothing; it remodels what is already present. Gould takes this as evidence that evolution is the product of a blind undirected process. But suboptimal design would also be a characteristic even of directed evolution, especially if the guiding intelligence, God, allows the play of chance in the process.

This also indicates that "design" need not be understood as a master plan that the evolutionary process executes in every detail. As Ian Barbour observes,

> If design is identified with the general growth toward complexity, life, and consciousness, then both law and chance can be part of the design. Disorder is sometimes the condition for the emergence of new forms of order, as in thermodynamic systems far from equilibrium, or in the mutations of evolutionary history.[42]

■ Directed Evolution and Self-Organization

As we have seen, many biologists believe that natural selection by itself is not enough to account for the increase of complexity over time. Also, it is not clear that natural selection can account for the origin of life. Another mechanism proposed for the origin of life and of evolution is "self-organization." What is this?

A simple example is the whirlpool that forms in a draining bathtub. Before the drain is opened, the water in the tub is quiet and at equilibrium, that is, it will stay in that state. Opening the drain produces a situation that is not at equilibrium; the state of the water is flowing through the drain. Energy is flowing through the system—the potential energy of the water is changing into kinetic energy as the water falls down the drain. And a novel form of order appears: the whirlpool, a self-organizing system that lasts only so long as there is water to feed it.

Classical science typically dealt with systems at equilibrium. The Russian-born Belgian chemist Ilya Prigogine won the Nobel prize for his investigation of self-organizing systems in far-from-equilibrium situations. He called these "dissipative systems" because they dissipate energy: energy has to flow through them for them to continue to exist (like the whirlpool in the tub). He described many examples of such systems. For instance, in his book *Order out of Chaos,* he describes a "chemical clock," in which a solution oscillates regularly from red to blue and blue to red. Given the initial mixture of reagents, the system is self-organizing.[43] Another example is the so-called "Lorenzian waterwheel." Imagine water falling into a waterwheel. If the flow is moderate and steady, the wheel turns at a steady rate. But if the flow is increased past a certain point, then the wheel may suddenly reverse direction, because the buckets do not have enough time to empty. And, in a little more time, it may reverse direction again, and again, and again, doing so in a cycle that is almost but not quite periodic.[44]

Now, although such things are relatively rare and exotic in the laboratory, they are common in life. Indeed, Prigogine thinks that life itself is the best example of a self-organizing, far-from-equilibrium, dissipative system (that is, energy has to flow through the system to maintain it). When the energy flow (i.e., food) ceases, the living being dies.

Prigogine discovered that often the change of state in a dissipative system may be due to a tiny fluctuation that becomes magnified and changes the whole system:

> A system far from equilibrium may be described as organized not because it realizes a plan alien to elementary activities, or transcending them, but . . . because the amplification of a microscopic fluctuation occurring at the "right moment" resulted in favoring one reaction path over a number of equally possible reaction paths. Under certain circumstances, therefore, the role played by individual behavior can be decisive.[45]

Prigogine himself relates this to the emergence of living systems, especially to the prebiotic phase of evolution, in which complex and possibly self-reproducing chemical systems emerged that (somehow) became the precursors of the first living things. In this process, tiny chance fluctuations may have been significant in steering the nascent process of evolution in one direction rather than in others.

Stuart Kauffman, a member of the Santa Fe Institute, has also argued that self-organization may be the key to the origin of life and to evolution. Working with computer models, he developed the idea of "auto-catalytic sets." Imagine, he says, 10,000 buttons scattered on a hardwood floor. Start connecting the buttons with threads: first connect one pair, then a separate pair, then another and so on. After a certain point (when the ratio of threads to buttons passes about 0.05), a giant cluster of connected buttons forms—a network. Now imagine that the buttons are chemical molecules, each of which is different, and each of which catalyzes a reaction between other molecules. As these molecules become linked together, they will catalyze an ongoing reaction that will be self-sustaining as long as there is energy being fed into the system (just like Prigogine's "dissipative structures"). Kauffman thinks that living things began in just this way, as autocatalytic sets of molecules.[46] Evolution generally, he thinks, cannot be explained by random mutation and natural selection alone, but must also be driven by processes of self-organization.

Kauffman's work has been criticized by biologists as too theoretical and computer-based. The question is: Would self-organization, or a combination of self-organization and selection, be enough to account for the enormous buildup of information that occurs in the process of evolution? First of all, "self-organization" is a descriptive term, not an explanatory one. It tells us that materials are self-organizing but does not explain how. Second, in using self-organization as an explanation for evolution, scientists are extrapolating from very simple, relatively undifferentiated systems like chemical clocks to enormously complex systems, like a cell. Cells, like all living things, have a hierarchical composition. They are made up of subunits, which are made up of subunits, and so on. An animal is composed of organs that are interrelated in physiological systems. These organs are themselves composed of tissues, which are composed of cells, which are themselves composed of subunits (the nucleus, mitochondria, etc.), which are composed of complex molecules, like proteins and DNA, which are composed of simpler molecules, etc. It may not be difficult for a homogeneous aggregate, in which the parts have no particular relation to one another, like hydrogen gas, water in a bathtub, or one of Prigogine's chemical clocks, to organize itself under certain conditions. But it is hard to see how an enormously complex system like a cell, articulated into subunits, *which cannot exist separately from the cell,* and which have unique and highly specific interrelations, could self-organize without the input of specifying information. (The same would be true for complicated machines like watches and computers.) Most biologists think that this information input can be explained by natural selection. I have given reasons for questioning that above. Self-organization may have a role in the organization of prebiotic life. But it is not clear that it could account for the buildup of information and specified complexity in something as intricately organized as a cell. But directed evolution would explain the development of such specified complexity.

◼ Directed or Undirected Evolution: Journey or Random Walk?

The bottom-line question, then, is: Is evolution directed in some way by an intelligence such as God, or is it undirected? Direction, to repeat, need not involve a detailed plan that controls every step of evolution. Few theologians today would want to defend such an idea. There is too much chance, waste (e.g., extinctions), and suffering in the evolutionary process. God's direction, then, if it exists, must allow for the chance, waste, and suffering that are intrinsic to the evolutionary process. Given this, a directed model of evolution fits the evidence at least as well as an undirected model of evolution, in which the whole process is taken to be the result of blind mechanical forces. If we take Denton's observations seriously, then it is hard to see how an undirected process could account for all aspects of evolution, though it could certainly account for some, such as the adaptation of bacteria to antibiotics or protective coloration in animals, which is explainable by natural selection. If the nematode is typical of organisms, then natural selection could explain microevolution. But it is hard to see how natural selection by itself could explain macroevolution, since the evolution of a species different from the nematode would seem to involve changing many genes simultaneously, in highly specific ways. Not only is it hard to envision how this could be done piecemeal, but self-organization would not seem to be a good analogy either. For self-organization in physics (a star), chemistry, or biology (slime molds) involves colonies of similar units whose relation one to another need not be highly specific. For a star, the individual atoms could be rearranged almost infinitely without changing the character of the whole. But this is emphatically not true of highly organized hierarchical systems like cells, organisms, or machines. Rearrange the parts of a watch or a computer and you have a dysfunctional pile of junk, not a new functional entity. There certainly are problems with models of directed evolution, but, as Denton notes, a successful theory does not need to explain everything; no theory does. It only needs to explain phenomena better than its competitors.

Directed evolution implies that there is a direction and hence a goal or end to evolution. There are several reasons for thinking that this might be the case. The first is the evidence and the arguments presented by Denton (above). The second is the apparent occurrence, in many experiments, of directed mutations. This by itself does not establish a case for directed evolution, but it undermines the case for random, undirected mutations, which have been and remain the strongest neo-Darwinian argument in favor of undirected evolution. A third reason is the continual increase in information and complexity over the whole span of evolution. Why should this be the case? Bacteria are very well adapted to their environments, probably even better than humans. There is no obvious reason why evolution should proceed to produce more and more complex beings over time. Spetner has showed that random mutation and natural selection by themselves cannot account for the buildup of information necessary for complex systems such as cells. A fourth reason is the evolution of humanity. Why should the evolutionary process produce intelligent creatures like human beings (and, for all we know, other kinds of intelligent creatures on other planets)? Human beings have many features that are not well explained by the neo-Darwinist theory of undirected evolution (see next chapter). These include free will, an intellectual capacity that goes far beyond the needs of simple survival, a moral sense that also seems to go beyond survival imperatives, and a religious sense. Why are human beings religious? The standard sociobiological explanation for religion is that it serves to strengthen the bonds of the individual to the group, and so to further survival. But this does not make sense. If this were all that was required, why would not evolution produce an instinct that drove persons to huddle together in groups for safety? (Indeed, we do observe such motives in people.) Why would evolution go to the expensive detour of leading persons to believe in God and an afterlife, and sacrifice large amounts of labor to make ornaments and weapons to be buried with the dead? Such a practice would put the group at a large competitive *disadvantage* against other groups that did not have religion, but which did have an instinct to stick together. And if it were at a disadvantage, it would be eliminated by natural selection. Thus the development of a religious sense in humanity seems to fit better with a model of directed evolution than with a model of undirected evolution. The same could be said of our intellectual capacity (which greatly exceeds what is necessary for survival), our moral sense, our sense of beauty, and our need for meaning and transcendence in life (these will be discussed in the next chapter). All of these seem like improbable products of a blind, random process, and are better explained as the result of a directed process.

Having said this, I think it unlikely that directed models of evolution will appeal to the scientific community, for a simple reason. Science is committed to naturalistic explanations. Any explanation involving an appeal to a transnatural intelligence is unlikely, in the present climate, to gain much hearing within the realms of science. But much of this is due to a bias in favor of metaphysical naturalism and materialism.

I think this is as much as we can say at present because, in my view, evolution is still a theory in the making. While evolution as such—that is, the development of more complex forms from simpler ones over time—seems well established, there is no consensus on the mechanism of evolution. Neo-Darwinist orthodoxy, that is, the explanation of evolution by random mutation and natural selection, still seems to be the dominant mode of explanation, but this also seems to be changing, especially with the advent of theories of complexity, chaos, and self-organization. If evolution is a theory that is itself in the process of change, it is risky to use it as a basis for a theology. The best we can do in such a case is to look at the most general and most well-established features of evolution, and try to construct a theology that takes these into account while leaving the details of the mechanism open. That is what I propose to do in the following pages.

■ A Theology of Evolution: Evolution as Journey

As we have seen, the feature of evolution that seems the most certain is that there has been a development from extremely simple, homogeneous systems to complex systems in which diverse elements are unified into a functioning system. This can be seen in the evolution of organisms, and humanity, but is also apparent in the evolution of the cosmos from the initial instant of the Big Bang to the present. Stars and galaxies are not enormously complex objects, but they are probably more complex than the pure energy/matter that constituted the universe just after the Big Bang.

Can this evolution toward complex systems be understood as providentially ordered by God?

Within a context of philosophical naturalism—that is, the worldview that only nature, and nothing greater than or other than nature exists—it is likely that evolution will be seen as a random walk, driven by chance and necessity, and human life as the fortuitous by-product of a blind process. There are at least two reasons for this.

First, natural science from the time of Galileo has rejected any final causes (that is, goals or purposes) in nature. Final causes are not quantifiable nor measurable, and so found no place in the "new science" of Galileo and Newton. Indeed, Jacques Monod declared that it is precisely the rejection of final cause that is constitutive of modern science.[47] Science instead concentrates on efficient and material causes. The method of modern natural science, then, simply could not disclose final causes, even if they were present. And this methodological limitation has become a metaphysical outlook: final causes don't exist, because science cannot perceive them, and we know that what science tells us is true.

Second, as Wolfhart Pannenberg reminds us in numerous writings, we cannot really know the whole of a process, including its aim or purpose (if there is any), until the process is complete. This is obvious in a journey; if a person sets out in an automobile from San Francisco, it would be foolish to try to guess the goal of her journey by the time she had reached St. Louis. Similarly, one cannot tell the end of a book from its first half. So even in principle it might not be possible to articulate the goal or purpose of evolution until the end of the evolutionary process.

Now Christian theology claims to know what that end is, but it is an end that can only be known through revelation: namely, the reconciliation of humanity and the cosmos with their Creator. The author of Ephesians writes: "[H]e [God] has made known to us the mystery of his will, according to his good pleasure that he set forth in Christ, as a plan for the fullness of time, to gather up all things in him, things in heaven and things on earth" (Eph. 1:9–10). This will be the kingdom of God, which was the principal object of Jesus' preaching and of his mission.

Furthermore, Christianity claims that the author of the whole universe and all life is God. This is clear in the first chapters of Genesis, but also in the first chapter of the Gospel of John, which says that "All things came into being through him" (i.e., the Logos). Pannenberg argues persuasively that the Logos, the second person of the trinity, is the ultimate principle of distinction, and hence the ultimate origin of distinct, created forms in the universe.[48]

Thus Christianity sees reality as structured according to a journey: the cosmos and its creatures come from God in creation, and are called back to fellowship with God in the eschaton, the end times. We can see this structure in the Gospel of John, in which the Logos descends into the world in the incarnation and returns to God in the resurrection. We can see it also in the *Summa Theologiae* of Thomas Aquinas, the first part of which treats the origins of creatures from God, and the last part of their return to God.

Indeed, traditional Christianity might even be said to have a kind of evolutionary structure, despite all the stereotypes of its being a "static" religion. The kingdom of God develops from a small seed (the parable of the mustard seed) to a great tree, and from a few disciples (a "founder population," to use Ernst Mayr's phrase[49]) to a movement spread across the earth. Furthermore, many of Jesus' sayings and parables emphasize that there is a selection process involved: "[M]any are called, but few are chosen" (Matt. 22:14).

Thus I think it likely that, when viewed within the context of a Christian worldview, evolution is likely to be perceived a kind of journey, which has, after all, a goal.[50] We can see in evolution a movement from simple systems to complex systems, both at the cosmic level and at the level of organic evolution, and we can also see this in the Christian view of history. For the end and culmination of history is, according to Christianity, the gathering of all the blessed into fellowship with God, through Christ and the Holy Spirit. And that is not all: Paul prophesies that the whole cosmos as well will "be set free from its bondage to decay and will obtain the freedom of the glory of the children of God" (Rom. 8:21). Certainly this would be the most complex system of all, containing maximal diversity but also maximal unity.

But to say that evolution, perceived within the context of Christian theology, is a journey is not to say that the sole aim of the evolutionary process is to produce human beings, or even intelligent life. Genesis 1 declares that all of creation is good in its own right, quite apart from the goodness of humanity. This is the point of the refrain, repeated after each day of creation: "And God saw that it was good." Thomas Aquinas also argues that God created so many and diverse kinds of creatures because only the whole panoply of creation would adequately express God's goodness. His reasoning is worth an extended quotation:

> For God brought things into being in order that his goodness might be communicated to creatures and be represented by them; and because His goodness could not be adequately represented by one creature alone, He produced many and diverse creatures, that what was wanting one in the representation of the divine goodness might be supplied by another. For goodness, which in God is simple and uniform, in creatures is manifold and divided; and hence the whole universe together participates the divine goodness more perfectly and represents it better than any single creature whatever.[51]

This point is also brought home by the doctrine of the resurrection, especially if that involves the cosmos, as Paul and the book of Revelation proclaim. As John Polkinghorne has written, and as traditional Christian doctrine affirms, in the eschaton, nothing that is good will be lost.[52]

But if evolution can be seen as a kind of journey, culminating in the reconciliation of all things in Christ, it is a journey fraught with tragedy. In nature, the price of life is the death of another, both to make room for new life and to provide resources for new life. This is true not only for individuals but for species; most species in the course of evolution have become extinct, and their death opened econiches that were then filled by new, creative forms of life. There is no better example of this than the extinction of the dinosaurs, who had flourished for more than a hundred million years. Their sudden extinction opened the way for the flourishing of the mammals, by the process biologists call adaptive radiation.

More than most religions, Christianity has resources to make sense of this tragic and paradoxical evolutionary history. For theologically, the tragedy, death, and subsequent creative transcendence of evolutionary history is the same pattern that is manifested in the life and death of Jesus: cross, death, and resurrection. The cross and death seem to be the necessary prelude to the transcendence of the resurrection, and it is a truism in Christian spirituality that one cannot ascend spiritually without first undergoing a dying to self. So in organic evolution, I see a pattern of cross, death, and subsequent transcendence.

Now, if creation comes from God, how does God act in the process of evolution?

First, of course, Christian theology has traditionally affirmed that God created the universe "from nothing." But we cannot conclude on scientific grounds that the Big Bang means that the universe was created from nothing. After all, there might have been a previous universe that imploded, or a previous condition, all trace of which has been lost. But at least we can say that the scientific scenario of the Big Bang is consistent with the belief that the universe was created from nothing. If we asked what a creation from nothing would look like scientifically, the answer is: like the Big Bang.

But God's creative work does not stop with the initial creation. To say that would be to adopt a modern form of Deism, in which God created the world like a watch, wound it up, and let it run on its own. Such a view is defended by Paul Davies (and many others), who argues that God designed the laws of the universe such that they would produce a system that developed by itself into complex living and intelligent beings.[53] Biologist Rudolf Brun puts it this way: "The fundamental insight of modern science that nature is capable of constructing itself is not yet fully appreciated by theology."[54] This might be characterized as a minimalist notion of directed evolution. But the biblical view is different: God's Spirit remains active as a creative principle in creation. As the psalmist declares in Psalm 104:30: "When you send forth your spirit, they are created; and you renew the face of the ground." God, then, sustains the universe in being, but is also active in guiding the process of evolution. This is known as God's *creatio continua* or "continuous creation."

How then might the Spirit act in evolution? I would agree with John Polkinghorne that the process by which God influences the creation is by the input of *active information*.[55] One way of thinking about God is as an infinite field of information, which transcends but also penetrates and sustains all entities. It is information that specifies form (see chapter 3). Each created, material entity is composed of energy with a specific form (electron, proton, photon, cosmic ray, etc.), located at a particular place in space and time. The particular form, place, and time of the entity is specified by information. It is, if you like, information crystallized into matter. God's sustaining of the created universe involves holding all energy in being, but also holding all created forms in being.

But at the quantum level, there is a certain amount of indeterminacy in elementary particles. Robert John Russell has proposed a model in which God acts at the quantum level to determine quantum indeterminacies, and so to affect some of the genetic mutations that in turn affect the development of species and the course of evolution.[56] Quantum events, unlike macroscopic events, are not entirely predetermined by previous causes. Rather, a quantum particle, such as an electron, exists in a range of possible states, which collapse into a single specific state when it is measured or interacts with another particle. In Russell's conception, God might act by causing the quantum state to precipitate in one direction rather than another, thus bringing about a mutation that would be favorable to evolutionary development. This would not involve an input of energy, only an input of specifying information, and so would not constitute a violation of any physical laws. But it would allow God to guide, albeit very gradually, the course of evolution.

The above model certainly allows for directed evolution, but has the weakness of implying that God only acts to determine some quantum events but not all. God therefore seems to act episodically. Russell is aware of this problem, however. He insists that God is omnipresent, that God continuously upholds all natural laws and all events, and so is continuously acting. But he wants to avoid the model that God determines every quantum event, because this would make God responsible for every event, including evil.

Perhaps the best analogue we have in trying to understand God's action in nature is God's action in human life. Believers assume that God sustains each of us in existence. But God also acts in special ways in the lives of persons, depending on the circumstances. Although God might be thought of as a field, this analogy is limited, because it does not do justice to God's personal qualities, such as responding to prayer.

God acts differently in the lives of different persons and in different circumstances. God's action in becoming incarnate in Jesus is different than his action in the lives of ordinary believers. So God acts uniformly in sustaining creation, but particularly in the lives of individuals. And the same can be said for God's action in evolution.

It is also true, I think, that God allows the created universe a great deal of its own freedom, so that, to an extent, it shares in its own ongoing creation. This is very similar to the relation of a parent and a child. The parent brings the child into being and thereafter can try to guide the child but cannot completely control its development. "Cannot" here is perhaps too strong; rather, the parent chooses not to coerce the child, but to let it develop on its own, while providing guidance and direction. This gift of freedom, as Polkinghorne notes, is the gift of love. If we really love the child, we will let her develop in her own way, choose her own vocation, and so on, while providing a supporting framework and guidance. To control the child completely would not be love, but dictatorship, and would not result in a child who can freely turn to the parent in love, but in a slave. So with God and creation. God brings it into being, including a framework of physical laws that allow nature to bring forth living and intelligent beings, but also allow it freedom, since his ultimate purpose is to bring forth free creatures who can know him and return to him in love. But at some points God influences the direction of evolution by influencing the information inherent in the system.

Niels Gregerson has argued that evolution is an autopoietic process, which is guided by God from within. Autopoietic means, roughly, self-forming or self-organizing. But, says Gregerson, God "influences particular processes by changing the overall probability pattern of evolving systems."[57] God, therefore, acts within the self-organizing capacities of nature to influence the direction of their evolution.

Another way that God might affect evolution is by whole-part influence (discussed in chapter 3). This idea has been particulary championed by Arthur Peacocke, who thinks God, as the whole context in which nature exists, can act to influence the behavior of the parts. Polkinghorne's example, cited in chapter 3, concerning the molecules of air in a room shows how sensitive physical processes can be to context. Even the motion of air molecules in a room is influenced by the gravitational force of matter on the other side of the universe. Practically speaking, then, even deterministic systems, like the successive collisions of molecules of air, are intrinsically unpredictable, even though the equations governing them are deterministic. This is one of the conclusions of chaos theory. Polkinghorne concludes from this that the processes of the universe are open, not closed, and that God, or the field of the Spirit, can act within them without "violating" physical laws.[58] Thus, if we think of God as the ultimate context of the universe, influencing information at the quantum level, it seems that God could affect sensitive systems, such as weather systems, and these in turn would influence the course of evolution.

As an example, let us consider a hypothetical instance of speciation. Most biologists now think that new species develop from so-called "founder populations," that is, a small fraction of a larger population that has become isolated from the parent population, perhaps because of a geographical barrier, like a stream, perhaps because of migration or some other cause. This founder population may have a gene reservoir that differs from that of the parent population. Furthermore, environmental pressures might push the small population in a different direction from its parent. Add to this the possibility of chaotic dynamics—that a small population may evolve rapidly in the direction of a different adaptation, and hence of a new species. Now, some of the influences catalyzing this change may be quite small—a slight change in weather patterns, or a change in the behavior of a few individuals, and so on. Such might be the pathways by which the Spirit could influence the progress of evolution.

It is important to note that none of these ways of God's influencing natural processes would necessarily be detectable by scientific methods. To a scientist, a mutation is random, and there is no way that it can be traced back to God. So also with the fluctuations in self-organizing systems, or with the factors that induce environmental change. Thus a theistically directed system of evolution may well be scientifically indistinguishable from an undirected evolutionary system. What the theist would explain by providence, the atheist or agnostic would explain as "chance."

Thus I think that God the Spirit, as the ultimate and yet immanent context of the whole universe, can certainly act to influence the course of evolution, not only by acts of intervention, but by the constant presence of a field that gently and imperceptibly draws creation toward an ultimate goal of reconciliation with its Creator. If the Spirit draws creation toward the ultimate future, and if that future is the reunion of all things with God, then this may be one explanation for the universal tendency toward increasing complexity and diversity in the universe.

I have another suggestion for how the Spirit might work in the world. We have seen in chapter 2 that ancient and medieval thinkers (such as Augustine and Aquinas) thought of the world as a hierarchy of being. In that worldview God was infinite, but there was still a unity between God and creation. Modern thought, however, tends to think of nature as an autonomous system, and God as outside of or extrinsic to nature. So if God were to act in nature, he would have to "intervene" from the outside, like a mechanic fixing a clock. God's transcendence was preserved, but God's immanence was lost. Yet for Aquinas, God is what is "innermost" in every entity. And God and creatures share a commonality in being, the simple act of existing. God exists by nature. God cannot not exist, but creatures exist by participation. They receive their existence from God. Therefore there is no absolute disjunction between God and creation, Spirit and matter. Creation and creatures proceed from the Spirit, and have their own independence. One instance of this independence is constant natural laws. But if there were an absolute disjunction between Spirit and matter, then there could be no influence from one to the other, in either direction. God could not influence events in our universe, nor could we influence God, through prayer or any other way. The idea that God is "wholly other" is problematic, for if it were true, we could know nothing of God, even that God exists. Nor could God reach us, even through revelation.

There is, then, a unity in being between God the Spirit and matter. But there is also a discontinuity. We and the physical universe are not little pieces of God—this is the view of pantheism. We are creatures possessing our own degree of autonomy. This is manifested in physical objects obeying the laws of nature, and in humans being able to make free choices, including the choice to reject God. I agree with Polkinghorne and others who hold that the gift of love is the gift by God of a degree of freedom and autonomy to creatures.

So, there is unity but also discontinuity. In Catholic thought, one way to speak of this is the language of participation. We can participate in the grace of the Spirit while yet being distinct from the Spirit. Similarly, I believe, the energy of the cosmos can participate in the creative field of the Spirit while being distinct from the Spirit. If there is a unity of Spirit and matter, and the cosmos is not a closed but an open system, then the Spirit should be able to influence matter from within, without "violating" the integrity of physical laws, probably by the input of information, perhaps at the quantum level.

Thus it seems reasonable, at least from the perspective of Christian theology, to believe that one factor in the course of evolution is the action of the Spirit, which like a field influences events ever so slightly, so as to move evolution, over millennia, toward its goal, which is the return of the cosmos to its Creator. One way for this to happen is for the cosmos to produce intelligent beings who can acknowledge their Creator in love and freedom. As Teilhard de Chardin saw, humanity is creation become conscious, and human worship is the creation returning its praise to the Creator.

What is the end of this evolutionary journey? I agree with those, like Teilhard de Chardin and Robert Wesson, who see evolution as moving from the biological level to the cultural level in the case of humans (and perhaps the higher primates). Furthermore, I would not rule out the possibility that evolutionary processes have produced beings of high intelligence elsewhere in the universe.

But in the Christian vision, the journey of humanity and indeed all of creation back to God is only completed in the resurrection. The resurrection of Jesus has traditionally been understood in Christianity as a promise of a more general resurrection, which will include all humanity and, if Paul is right in Romans 8, even the creation itself. This is of course deeply mysterious, since we have no examples of it in our present world. Furthermore, is it even conceivable that biological and cultural evolution could evolve to a kind of spiritual evolution culminating in the resurrection? Let me offer a few reflections.

First, and most obviously, the resurrection seems to involve a different kind of materiality from our own. Jesus' body, before his ascension, was palpable to his disciples—clearly physical then, but also seemingly not constrained by the limitations of space and time (see John 20:26). A Christian notion of the resurrection is not a resuscitation; the resurrected body would not be just like our bodies now, but would be, as Paul puts it in 1 Corinthians 15, a spiritual body. There is a continuity with present materiality, but also a discontinuity. The resurrection involves a kind of transcendence for which we have no evidence besides the resurrection of Jesus himself.

The only way this is possible is through the agency of the Spirit. If matter and Spirit are, as I have argued, continuous in some sense, then the kind of transcendence imaged in Jesus' resurrection might be a possibility. The resurrection does not involve a "jump" to nonmateriality, but it does apparently involve a transcendence to another kind of materiality.

But if such a transcendence is possible, I do not see it as a gradual transcendence. Jesus' resurrection was preceded by a terrible death. We have seen that this pattern seems to be part of the story of evolution: cross, death, then transcendence. I would expect the same to be true in the case of the resurrection. As Polkinghorne holds, the new creation is consequent upon the death of the old.[59]

As a Christian, I see the resurrection as the reconciliation of all humans, and all creation, with Christ, the incarnate Logos, through whom the creation was formed. This is the vision of the kingdom of God, the object of Jesus' preaching, and the vision of Paul's letters and of the book of Revelation. It would seem to be the ultimate in unified diversity, or in other words, the ultimate in complexity. It is tempting to see in this icon a kind of dynamic attractor, to use the language of chaotic dynamics: a stable pattern that draws the long process of evolution toward a greater harmonic diversity and complexity, to its final consummation. I believe this is also consonant with the vision of Pannenberg: that the Spirit draws the process toward a goal that, like any journey, cannot be fully understood except from the perspective of the end.

6

Human Nature

Do We Have Souls?

■ The Big Bang theory and the theory of evolution may seem remote from the concerns of everyday life. After all, the average person can get through life, run a business, or be a successful professional while knowing little about the origin of the universe or the evolution of life. As Dr. Watson recounted in *A Study in Scarlet,* Sherlock Holmes did not know that the earth revolved around the sun, and when he was told that it did, said he would do his best to forget it, since it had no bearing on his work.

But this is not the case with theories about the human person. Our concepts about human origins, nature, and destiny are critical for our conceptions of who we are and what we should do in this life. So it is here that philosophy and theology really matter. If we see ourselves as mere collections of molecules, the accidental spin-offs of a random process, with no intrinsic meaning or purpose in life, and no hope of afterlife, a negative self-concept is likely (imagine teaching this to kids). If we think that we are children of God, possessing intrinsic dignity, free will, and intellect and created for a transcendent destiny and purpose, that will give us a more positive self-concept. The legal and political institutions of the West have been based on the latter conception of humanity. Our legal system, based as it is on the notion of personal responsibility, presupposes the reality of free choice. We are not determined in our choices and actions, but free. In a given situation, we can choose several possible responses; we are not programmed to follow one course of action. But it is hard to make a case for free choice in a naturalistic view of the human person. Increasingly, those philosophers who have adopted a naturalistic perspective do not defend free will, but regard it as an illusion. So also our political institutions presume the reality of intrinsic or inalienable rights, which are God-given and therefore cannot be denied by the state. But again, it is hard to see how or why we should assert that persons have inalienable rights if they are no more than products of natural processes. In that case, they are ontologically equivalent to animals, only more clever. But we do not accord inalienable rights to animals. Rather, we cull out the weak and preserve the healthy. There is no reason to think that we would do otherwise if a fully naturalistic view of humanity prevails in our civilization.

The subject of the human person involves many disciplines: biology, psychology, sociology, anthropology, philosophy, theology, and others. Unavoidably, the topic is complex. Therefore, it will be necessary to present summaries of positions and refer the reader to more extended discussions in the notes.

In spite of the complexity of this subject, it is possible to distinguish four broad perspectives on the human person. First, most religions, including tribal religions (e.g., traditional African and Native American religions), Hinduism, and Christianity, have traditionally held that the animating force in a person is a soul, which is nonmaterial and survives bodily death. The second position also believes in an immortal soul, but places its main hope in the resurrection of the body. Traditional Judaism, Christianity, and Islam are examples (these religions traditionally affirmed both an immortal soul and resurrection). A third position, naturalism, holds that human beings evolved and can be completely explained by natural processes. Humans are psychosomatic unities, and there is nothing in the person that cannot be explained by physics, chemistry, and biology. Persons have no souls, nor is there a possibility of resurrection. Free will is also suspect, since it is hard to explain it naturalistically. The fourth position, emergentism, holds that persons do evolve through natural processes, and that there is nothing in them other than molecules. But it also holds that complex systems, including the human brain, develop new properties, which are properties of the whole, but not of the parts. These properties, such as life and consciousness, emerge as the parts (molecules) come together to form complex wholes (such as cells, animals, and persons). Those who argue this typically would deny that there is a soul which survives bodily death, but might defend a resurrection, in which the whole person, a psychosomatic unity, is restored in a new environment. Most of those in the theology-science discussions would hold some form of this position.

Most persons would hold positions one or two. That is, they would believe in a soul that survives bodily death. This is the traditional teaching of tribal religions, of the major Eastern religions (Hinduism and some forms of Buddhism), as well as of the traditional monotheistic faiths: Judaism, Christianity, and Islam. But very few in the sciences would support either position one or two. Thoroughgoing naturalists, who deny any supernatural realty, would hold position three. And many, probably most, theists who are also scientists, or are involved in the science-theology discussions, would affirm some form of emergentism. The strength of emergentism is that it allows one to affirm the scientific consensus that the human person is a psychosomatic unity, while still holding onto Christian (or other theistic) beliefs about free will, creation in the image of God, and resurrection. Thus this is an attractive position for those who wish to reconcile contemporary science with religion.

However, for reasons to be discussed below, I will defend the traditional Christian teaching that the human being is a unity of two distinct principles: a soul that can survive bodily death, and the matter of the body. I will not, however, defend the dualism of Descartes (body and soul are independent substances). Rather, I will argue for what might be called a dipolar monism, in which the soul is the organizing principle of the body, but can survive bodily death. In this view, then, science and theology are complementary. Both are needed to arrive at a full truth of the human person. Systems that are too spiritual and ignore bodily constraints and systems that are too materialistic and ignore spiritual realities, such as free will and the soul, both fail to reach a comprehensive view of the human person. The chapter will present a brief history of pertinent philosophical and theological ideas about the human person, and then consider the reductionistic naturalist account of the person and its deficiencies. The following chapter will take up emergentist accounts, consider some criticisms of these positions, and conclude with a case for a traditionalist Christian account, that the person is best understood as an embodied soul, conditioned both by the body and by social relations.

■ Biblical Thought

Ancient Israel

According to Genesis 2, God created Adam from the dust of the ground, and breathed into his face the breath of life. This would seem to indicate that in Hebrew thought, human beings were made up of two components: matter and spirit. But later Hebrew writing, for example, the book of Daniel, presents the hope of the bodily resurrection, which would seem to indicate that the Hebrews thought of the human being as a totality, a psychosomatic unity. There was no separated soul to carry the personality after death, so the argument goes; therefore the only hope for immortality was the resurrection of the whole person. For in Hebrew thinking, there could be no person without the body.

Now, this latter opinion, though foreign to the thinking of an earlier generation of biblical scholars, such as Helmer Ringgren,[1] has become virtually a consensus among recent scholars of the Hebrew Bible. James Barr, in *The Garden of Eden and the Hope for Immortality*,[2] traces the rise of this thinking among Hebrew Bible scholars. It began with the "dialectical theology" of the 1920s, and gained impetus with the publication of Oscar Cullmann's book *Immortality of the Soul: or, Resurrection of the Dead?: The Witness of the New Testament*.[3] Cullmann argued that the belief in an immortal soul and the belief in the resurrection were opposed. The latter was a Hebrew idea, while the former was a Greek idea, which had been projected back into the Hebrew Bible and the New Testament by Christian scholars. But according to Cullmann, the anthropology of both the Old and New Testaments is not dualist, but monist: the human being is a psychosomatic unity. Thus Harvard biblical scholar Krister Stendahl echoed Cullmann's conclusion: "The whole world that comes to us, through the Bible, OT and NT, is not interested in the immortality of the soul."[4] And Scripture scholar Joel Green writes: "It is axiomatic in Old Testament scholarship today that human beings must be understood in their fully integrated, embodied existence. Humans do not possess a body and soul, but are human only as body and soul."[5] One finds this opinion expressed over and over again in science and theology literature.

But this opinion ignores certain critical texts. For example, in 1 Samuel 28, Saul consults a medium in Endor. The woman summons up the spirit of Samuel, who rises out of the ground and demands: "Why have you disturbed me by bringing me up?" (1 Sam. 28:15). There is no doubt that it is Samuel who addresses Saul, and there is no doubt that Samuel is dead, but has been lingering in the underworld. While it certainly is true that one could not be fully human without the body, the passage in 1 Samuel indicates that at least the personality can survive bodily death. Samuel speaks cogently to Saul, and even predicts what will happen to Saul on the next day. Yet Samuel has not been resurrected.

Furthermore, necromancy is mentioned as being practiced in Israel in Isaiah 8:19 and 2 Kings 21:6, and it was forbidden by written law (Lev. 19:31; 20:6, 27; Deut. 18:10–11). Needless to say, one does not forbid by law what no one practices anyway. If there were proscriptions, there must have been practices against which the laws were addressed.

Finally, the wider canon of the Hebrew Scriptures, which included later writings such as the books of Maccabees and Wisdom of Solomon, clearly referred to the immortality of the soul. Thus we read in Wisdom 3:1–3: "But the souls of the righteous are in the hand of God, and no torment will ever touch them. In the eyes of the foolish they seemed to have died, and their departure was thought to be a disaster, . . . but they are at peace." This reading is common in Catholic funerals. But since Wisdom, like Maccabees, is not found in the Protestant Bible, Protestant biblical scholars have not paid much attention to it.

Old Testament scholar James Barr considers this problem of "Hebrew 'totality thinking' and the soul" at length in *The Garden of Eden and the Hope for Immortality*. Since he is opposing a majority of biblical scholars, I will cite him extensively. He writes:

Is it even remotely plausible that ancient Hebrews, at the very earliest stage of their tradition, already had a picture of humanity which agreed so well with the modern esteem for a psychosomatic unity? . . . Is there not an obvious bias in so many modern textbooks, which seem to want nothing more desperately than to deny that the Hebrews had any idea of an independent "soul," worse still an immortal one? . . . It is obvious that the enormous modern cultural pressure in favor of an understanding as the "totality" of the person has moved scholarship in that direction.[6]

He concludes:

I submit, then, that it seems probable that in certain contexts the *nephesh* is *not*, as much present opinion favors, a unity of body and soul. . . . [I]t is rather, in these contexts, a superior controlling center which accompanies, expresses, and directs the existence of that totality, and one which, especially, provides life to the whole. . . . It is *particularly* the thought of the envelopment of the *nephesh*, in the sense of "soul," in Sheol that is hateful. With the recognition of this fact the gate to immortality is open. . . .

In spite, therefore, of the importance of Hebrew totality thinking—which I do not dispute in principle—some formulations of the Hebrew Bible point towards more dualistic conceptions. As we have seen, God's formation of Adam, "dust from the earth" and his breathing "into his nostrils the breath of life" (Genesis 2:7) . . . look awfully like two ingredients, of which Adam was a sort of compound. . . . Hebrew thinking, then, is not itself a monolithic block. . . . The accepted contrast between a Platonic philosophical picture and a Hebrew totality concept may be constructed upon very limited sectors and periods on both sides, arbitrarily selected and inadequately investigated.[7]

Thus, it is not true that Hebrew conceptions of the human person neatly coincided with the modern fashionable opinion that the person is a psychosomatic unity. Some conceptions indicate that the core of the personality, as in the case of Samuel, could survive bodily death, even in Sheol. (This is very similar to the thought expressed in Homer's *Odyssey,* book 11.) Some conceptions, such as that found in Wisdom (a late writing influenced by Greek Platonism), point clearly to a notion of an immortal soul. All of these various beliefs carried forward into the world of New Testament thought.[8]

The New Testament

New Testament anthropology is quite complex. Any adequate treatment would have to deal with the perspectives of different authors: John's anthropology, Paul's anthropology, the view of the Synoptic Gospels, and so on. There is not space for that here. Rather, I will focus on the claim, almost universal in the science and theology literature, that the New Testament view is that the human person is a psychosomatic unity, and that the New Testament does not teach the "dualist" view that the person is a soul and a body. I maintain that there are diverse views in the New Testament. Many New Testament writings support the traditional Christian view that the human person is resurrected after death. But many also indicate that there is a core of the personality, a soul that survives bodily death. Rather than these teachings being in competition, as many commentators would have us believe, I hold that they are complementary. For it is the soul that provides the personal continuity between the person who dies and the person who is resurrected. Thus the best interpretation of the New Testament is that the person is an embodied soul, who will be resurrected in the end times.

We might start with a statement of Oscar Cullmann that "The Jewish and Christian interpretation of creation excludes the whole Greek dualism of body and soul."[9] This is cited by Ian Barbour as supporting his view that the person is "an embodied self, not a body-soul dualism."[10] Barbour also cites author Lynn de Silva as follows:

> Biblical scholarship has established quite conclusively that there is no dichotomous concept of man in the Bible, such as is found in Greek and Hindu thought. The biblical view of man is holistic, not dualistic. The notion of the soul as an immortal entity which enters the body at birth and leaves it at death is quite foreign to the biblical view of man. The biblical view is that man is a unity; he is a unity of soul, body, flesh, mind, etc., all together constituting the whole man.[11]

Now, first of all, careful reading of biblical anthropology will show that there is not just one view of humanity in the Bible. There is a diversity of views, as Barr points out in the passages cited above. So a blanket generalization, such as that of de Silva, ought to be suspect. Furthermore, as Barr also indicates, the antidualism ideology is a result of cultural pressure. Dualism is a politically incorrect view in academia. It is associated with hierarchy, patriarchy, and oppression, and furthermore cannot be substantiated scientifically, for science cannot find any trace of a soul. Furthermore, "dualism" in today's parlance almost always means the dualism of René Descartes (see below), which I agree is indefensible. It is this kind of dualism that is projected back onto the Christian tradition, and then rejected. But, as I will explain below, the Christian tradition never held the dualism of Descartes until the seventeenth century. The "dualism" of Aquinas was very different, and probably should not be called "dualism" at all.

Barbour's view, that the person is an embodied self, a psychophysical unity, not a body-soul dualism, is held by most writers publishing in the area of science and theology. However, there are many passages in the New Testament that witness to the belief in a soul that survives bodily death. New Testament authors also, of course, teach the resurrection of the body. But the only reason that the resurrection of the body and the belief in an immortal soul would be seen as contradictory would be if we assume a Cartesian-style dualism. But if we think of the soul as the form of the body, as Aquinas did, then an immortal soul and a resurrected body can be understood as complementary. So in this section, I will consider New Testament passages, from multiple authors, that support a separated soul. Later I will consider how this can be reconciled with the view that the person is a psychosomatic unity.

Let us begin with some of Jesus' statements as reported in the Synoptic Gospels.[12] One is Matthew 10:28: "Do not fear those who kill the body (*soma*) but cannot kill the soul (*psyche*); rather fear him who can destroy both soul and body in hell." The parable of the rich man and Lazarus (Luke 16:20–25), in which the deceased rich man in hell sees the deceased Lazarus in Abraham's bosom, presupposes the survival of their personalities (and therefore souls) after death. Jesus' words to the dying thief on the cross—"Truly I tell you, today you will be with me in Paradise" (Luke 23:43)—also seem to imply personal survival after bodily death.

Even Paul, whose anthropology is often taken as the best evidence in the New Testament for a unitary anthropology, wrote: "I know a person in Christ who fourteen years ago was caught up to the third heaven—whether in the body or out of the body I do not know; God knows" (2 Cor. 12:2). This "person" is generally taken to be Paul himself. But if he can speak of the possibility of being caught up to heaven "out of the body," he must think it possible for the personality to exist outside of the body. This is also implied in his comparison of the body to an "earthly tent": "[W]e know that if the earthly tent we live in is destroyed, we have a building from God, a house not made with hands, eternal in the heavens" (2 Cor. 5:1). Barr remarks: "It looks to me as if Paul had no idea that he was supposed to be working with Hebraic totality concepts or any other such thing."[13] Further evidence for belief in a soul is found in non-Pauline letters. First Peter 3:18–19 reads "For Christ also suffered for sins once for all, the righteous for the unrighteous, in order to bring you to God. He was put to death in the flesh, but made alive in the spirit, in which also he went and made a proclamation to the spirits (*pneuma*) in prison."

Finally, several passages in the book of Revelation (a late text, but still expressive of the New Testament worldview) refer to the souls in heaven of those who have been martyred, but not yet resurrected. One such is Revelation 6:9: "When he opened the fifth seal, I saw under the altar the souls (*psyche*) of those who had been slaughtered for the word of God and for the testimony they had given; they cried out with a loud voice, "Sovereign Lord, . . . how long will it be before you judge and avenge our blood . . . ?"

Why, then, do so many contemporary scholars hold that the New Testament does not teach the immortality of the soul? Barr claims that "much of the turn against immortality of the soul was not a return to the fountainhead of biblical evidence, but a climbing on the bandwagon of modern progress."[14]

Certainly the New Testament also teaches the resurrection of the body. No one disputes this. It is found in Jesus' teachings (cf. Matt. 22:23–33 and parallel passages), in Paul (1 Cor. 15), and elsewhere. But to argue that this opposes belief in an immortal soul does not make sense. In fact there is a diversity of views in the New Testament. I conclude with the words of James Barr:

> What is not acceptable is the attempt, in the name of a dubiously constructed picture of Hebrew thought, to override rather obvious meanings of the Greek New Testament. The New Testament certainly says little directly and specifically about the immortality of the soul; but it has a reasonable degree of mention of immortality, and it certainly has an awareness that things of the body and things of the soul could take different directions.[15]

■ Christian Tradition

Both the postbiblical Jewish and Christian traditions teach the immortality of the soul and the resurrection of the body. In the article on "Afterlife" in the *Encyclopedia Judaica* we find the following:

> When a man dies, his soul leaves his body, but for the first twelve months it retains a temporary relationship to it, coming and going until the body has disintegrated. Thus the prophet Samuel was able to be raised from the dead within the first year of his demise. This year remains a purgatorial period for the soul, or according to another view only for the wicked soul, after which the righteous go to paradise and the wicked to hell. . . . In Maimonides . . . the immortality of the soul is paramount. Though he makes the belief in resurrection, rather than in the immortality of the disembodied soul, one of his fundamental principles of the Jewish faith . . . it is only the latter that has meaning in terms of his philosophical system.[16]

Though there is not a lot written about it, the survival of the soul after death seems to have been presumed in the early Christian tradition. Justin Martyr in his *Dialogue with Trypho* presents the belief that after death the souls of the righteous go to some "better place," and the souls of the wicked to some "worse place," to await judgment. Justin makes the important point that the soul is not naturally immortal (as it seems to be in Platonism). Rather, it is God who gives it life: "the soul shares in life, when God wants it to live."[17] If God withdraws its life, the soul ceases to exist.

The substantial nature and immortality of the soul is articulated more fully in the work of St. Augustine. The human being, he says, is a "rational soul using a body."[18] Elsewhere he states that "A soul having a body does not make two persons but one man (*homo*)."[19] The powers of reason and understanding are present in the soul from infancy, and are awakened and developed as the child ages.[20] Because Augustine is convinced of the immortality of the soul, he does think that some souls undergo purificatory punishment after death. "Temporary punishments are suffered by some in this life only, by others after death, by others both now and then, but all of them before that last and strictest judgment."[21]

Augustine wrestled all his life with the problem of the origin of the soul, but never reached any definite conclusion. The soul, he thought, was intermediary between God and the body. "The metaphysical reason for the union of the soul and body in Augustine is simply that the soul is to serve as an intermediary between the body it animates and the Ideas of God that animate the soul."[22] Augustine did not follow Socrates' idea, taken up by Origen, that the body was the prison of the soul. Everything God made is good, including the body. In Eden, there was no suffering; death was the result of Adam's sin. The corruptible body was a load on the soul, as written in the book of Wisdom (9:15), but that is the result of the sin of Adam: "The soul is weighed down not by the body as such, but by the body such as it has become as a consequence of sin and its punishment."[23] The consequence of sin is that the soul has lost its mastery over the body, because it has turned from God.

For Augustine, the effects of original sin have been catastrophic. The whole human race is implicated in Adam's sin, and so is a *massa damnata:* "Hence the whole mass of the human race is condemned; for he who at first gave entrance to sin has been punished with all his posterity who were in him as in a root, so that no one is exempt from this just and due punishment, unless delivered by mercy and undeserved grace."[24] A consequence of this belief is that, for Augustine, no one can be saved outside the church and its sacraments. Unbaptized infants will burn in hell, a repulsive doctrine that discomfited Augustine, but which followed inexorably from the logic of his position. This very dark view of the human situation cast long shadows across the history of Western thought, influencing Aquinas and the Catholic tradition, but also Luther, Calvin, and the Protestant tradition.

Augustine never solved the problem of how the soul was related to the body. The soul, he thought, was a substance, yet the body was also a substance. And the human being was a single composite substance. He realized that this caused philosophical problems, but he was unable to resolve them.

Thomas Aquinas followed Augustine in many respects, but he approached the unity of the human person from an Aristotelian rather than a Platonic perspective. Thomas argued that the soul was the form of the body, in the Aristotelian sense of form, but was also able to exist on its own, apart from the body. It was, then, a subsistent form. The body was not itself a substance, for without the form informing the body, it would have no form or organization of its own. Thus Aquinas was able to assert that the human person was a unitary substance, a unitary being, composed of two principles: form and the (prime) matter of the body. Neither a soul using a body, nor a soul in a body; the person was a soul-body composite in which the matter of the body was "formed" or organized by the soul.

Aquinas's notion of form here is not that the form is simply a static pattern or organization, like that of a chemical molecule. Rather, the form is a dynamic internal organizing principle, which contains within it a final end or goal, which the organism strives to fulfill. In the case of a human, this intrinsic end or goal was to know and love God. Human beings, therefore, are created for a more than natural end, fellowship with God in heaven.

Interestingly enough, Aquinas did not think that the human soul was created at conception. He thought that the developing embryo had first a simple plant soul, then an animal soul, and only in the last months, after the brain had been formed, a fully human soul. (Aquinas did not, however, approve of abortion for this reason. He argued that it was God's will that an embryo become a human person, and that to kill it even before it had a human soul was a grave sin.)

■ Protestant Thought

While it is not surprising that Roman Catholicism, with its veneration of the saints and doctrine of purgatory, would be committed to the immortality of the soul, it is also true that Eastern Orthodoxy and classical Protestantism were firmly committed to immortality as well. Though Lutheran confessional documents say little about the state after death, Luther does suggest, in the Smalcald articles, that the saints in heaven might pray for us (article 2). John Calvin also teaches that the soul is immortal. "Moreover, there can be no question that man consists of a body and a soul; meaning by a soul, an immortal though created essence, which is his nobler part."[25] The Westminister Confession, following Calvin, affirms that the soul is immortal, is judged immediately upon death, and goes to heaven or hell, there to await the resurrection of the body. Indeed, as Barr reminds us, throughout the Jewish and Christian tradition the doctrine of the immortality of the soul and the resurrection of the body were regarded as complementary.[26] It was the soul that provided personal continuity, and the soul that received the body restored by the resurrection. If, as in Catholic thinking, the soul was the form and activating principle of the body, then a resurrection in the absence of the soul would be an absurdity and an impossibility.

■ Modern Conceptions of Human Nature

The modern conception of the human person is usually said to begin with René Descartes. This is because Descartes rejected the Aristotelian and Scholastic theory of form and final cause, and accepted instead the new atomic and mechanistic theory of matter, which is at the heart of modern science. Within that context, he tried to work out a philosophy of the person and the mind. The body, he thought, was governed by simple mechanical principles. An example of this was the reflex. If you tap the knee at a certain point, it will jerk. This action is deterministic (not freely willed), predictable, and hence mechanical. This idea could be expanded to explain most actions of the body.

But Descartes also understood that many of our actions are not initiated by mechanical causes; rather, they are initiated by free decisions. I can decide to move my knee. In this the ultimate cause of the action is my free choice. So Descartes concluded that the mind is not governed by mechanical or physical principles. The mind, he thought, was an immaterial substance, free and immortal. His conviction was based on several philosophical arguments. For example, he argued that I can doubt everything, even that I have a body. But I cannot doubt that I think, because the act of doubting presupposes the act of thinking. Therefore, in my essence, I am a thinking being. But thinking cannot be explained by mechanical principles like reflexes. Therefore my mind is immaterial. I am composed of both a mechanical body and a nonmaterial mind, which thinks, chooses, doubts, and decides. There are two substances, then, in the human being: the material body, which, like all material objects, is characterized by extension in space, a *res extensa* (extended thing), and a mind, which is not extended in space, but which thinks, a *res cogitans* (thinking thing). This is Descartes's idea of mind-body dualism: dualism because there are two independent substances in the human being, not one. In this way, then, Descartes was able to affirm the new sciences, which explained matter on mechanical principles, but also preserve the traditional Christian teachings about free will, the immateriality of the mind and soul, and the immortality of the soul.

The great problem faced by Descartes's philosophy (as by all forms of dualism) was how these two substances could interact. How could the immaterial mind affect the material body? Descartes thought that the connection lay in the pineal gland. Somehow, the mind affected the pineal gland, which in turn affected the body, so that a purely mental decision, like deciding to move one's arm, could be translated into mechanical action. Later followers of Descartes proposed different solutions to this problem. One of the most curious was that of Nicholas Malebranche, who thought that the only connection between the soul and the body was God. When the soul decided to do something, God caused the body to do it. Every occasion was caused by God, hence this doctrine is known as "occasionalism."

Descartes's dualism, which separated the mind and the body,[27] lasted down to the late twentieth century. The philosophers of the Enlightenment typically thought of the human mind as free, but the body as controlled by mechanical principles. But several factors led to a dramatic loss of the credibility of dualism in the last century. First, the rise of an evolutionary explanation for human origins has made it harder to explain the origin of an immaterial mind and soul. Human beings seem to emerge by degrees from primate ancestors. Thus the overwhelming tendency in contemporary biological thought is to see human beings as different from animals only in degree, not in kind. But an immaterial mind would imply that humans are different from animals in kind, not just in degree, since almost no one believes that animals have immaterial minds. Second, modern neuroscience has strongly reinforced the scientific conviction that the mind has its roots in the brain. Damage to the brain, as occurs in Alzheimer's disease, has severe repercussions on the mind. Drugs (e.g., LSD), even in minute quantities, can drastically alter the workings of the mind. So it seems self-evident to those in the neurosciences that the mind is an aspect of the brain and could not survive without the physical brain. Third, several currents of modern thought, both in philosophy and the sciences, emphasize that the human person is not an isolated substance, a radical individual, but emerges within a social matrix and is an inherently social being. Human consciousness develops only through social interaction, and without social interaction remains truncated.

Thus the picture that emerged at the end of the twentieth century is that of the person as a psychosomatic unity existing within a dynamic matrix of social relationships. The intellectual currents have swung sharply against the ancient and medieval belief in the person as composed of an immortal soul and a body, and against Descartes's notion of the person as an individual immaterial substance, a mind in a body. Let us consider in more detail two of the main reasons for this: the modern theory of evolution and the development of modern neuroscience.

■ The Evolution of Humanity

Probably the first attempt to apply evolutionary theory to the emergence of humanity was Charles Darwin's *The Descent of Man and Selection in Relation to Sex*, published in 1871.[28] Since then, there is an entire field—paleoanthropology—devoted to explaining evolutionary human origins. And evolutionary explanations for human behavior, morals, and religion are in vogue among evolutionary psychologists and sociobiologists. In the following brief sketch of the evolution of humanity, I will follow the account of Ian Tattersall's book *Becoming Human*.

Tattersall thinks that the origins of hominization—the evolution of humans—began about 4.4 million years ago, when some African primates began to adopt a bipedal, upright posture. These primates lived in areas where forests bordered on grassy savanna. In the forest they would brachiate through the trees, but in the savanna they walked upright, at least for short periods. There are several theories as to why this happened. Perhaps an erect posture gave these early hominids an advantage in detecting predators, such as lions, leopards, and hyenas, because they could see over the high grass while foraging for food. Perhaps it was a way of regulating body heat, since less of their bodies would have been exposed to the tropical sun. Probably it was a gradual process. But it conferred certain evolutionary advantages. It allowed them to scan for predators. It freed the hands, so that simple tools like digging sticks, or simple weapons, could be made and carried. Perhaps, eventually, sign language developed. An early example of such a hominid was *Australopithecus afarensis*, of which "Lucy" is the most famous member. We should not think of these early hominids as human, however; they were probably more like "bipedal apes."

For the next 1.5 million years there was no great change in early hominid lifestyle. Then, about 2.5 million years ago, the first tools appear in the archaeological record. This is probably correlated with a climate shift in Africa; a more arid climate led to the conversion of forests and woodlands to savanna. *Homo habilis*, man the toolmaker, appears at about this time. The tools themselves were very simple: sharp stone flakes sheared off from stone cobbles by using a stone hammer. These tools were carried around, so that stone-cutting flakes could be manufactured when and where they were needed, namely, to butcher a freshly killed animal.

Again, there is a long period of stasis. But about 1.5 million years ago, large stone hand axes appear in the archaeological record, indicating an advance in stone working. This appears about the time of *Homo ergaster* in Africa. Along with this appears the use of fire.

About 50,000 years ago, *Homo heidelbergensis* appeared in Europe. This species was possibly the ancestor of Neanderthal man and also of ourselves. With this species, the first recorded use of shelters appears. Its brain volume was about 1,100–1,200 cubic centimeters, approaching that of modern man.

One of the species that succeeded *Homo heidelbergensis* was the Neanderthals—*Homo neanderthalensis*. This species flourished in Europe and west Asia from about 200,000 years ago to about 30,000 years ago, when it became extinct. It would have shared territory with our own species, *Homo sapiens*, for about 60,000 years. Indeed, it probably came to be extinct because it could not compete successfully with *Homo sapiens*. Neanderthal brains were as large as ours, but were structured differently. They lacked the large frontal lobes of the neocortex, which is where most of the thinking is done by modern hominids. Consequently, they lacked the large forehead characteristic of *Homo sapiens*. Rather, their head swept back just above the eyebrows, but protruded at the back. They perfected the shaping and use of stone tools and axes, but never seem to have progressed beyond the use of stone tools. Tattersall doubts that they had the use of language. There is evidence that they did occasionally bury their dead, but there is no unambiguous evidence of grave goods being buried with the dead. Similarly, there is no real evidence of artistic expression among the Neanderthals. Tattersall concludes that "symbolic expression was not a regular aspect of Neanderthal life."[29]

Homo sapiens seems to have emerged in Africa about 120,000 years ago. The first unambiguous evidence of modern human behavior comes later. Tattersall detects artistic activity in Africa about 60,000 years ago. This indicates the development of symbolic thought, and may occur at the same time as the development of language. Tattersall, along with many others, thinks that what is most distinctive about modern humans is the use of symbolic thought, that is: "The ability to generate complex mental symbols and to manipulate them into new combinations. This is the very foundation of imagination and creativity: of the unique ability of humans to create a world in the mind and to re-create it in the real world outside themselves."[30] Symbolic thought is epitomized in language, for languages assign certain sounds to represent objects, actions, and concepts. The sounds chosen are not onomatopoeic; they do not evoke the object, as a cry or a pantomime represents what it expresses. Sounds as various as "house," "casa," or "domus" can represent the same thing. And further, abstract ideas can be represented in language. Like language, art and ritual can also indicate the presence of symbolic thinking. Thus the wonderful Cro-Magnon cave paintings, found in Lascaux Cave, France, and elsewhere, indicate a high degree of artistic sensitivity.

Another distinctive feature of human behavior that emerges with Cro-Magnon humans is elaborate burial rituals, which strongly suggest a belief in afterlife. Tattersall describes a striking instance. At the 28,000-year-old site of Sungir, in Russia, three individuals were found buried and "dressed in clothing onto which more than three thousand ivory beads had been sewn; and experiments have shown that each bead had taken an hour to make. They also wore carved pendants, bracelets, and shell necklaces." He continues:

> . . . In all human societies known to practice it, burial of the dead with grave goods . . . indicates a belief in afterlife: the goods are there because they will be useful to the deceased in the future. . . . What is certain is that the knowledge of inevitable death and spiritual awareness are closely linked, and in Cro-Magnon burial there is abundant inferential evidence

for both. It is here that we have the most ancient incontrovertible evidence for the existence of religious experience.[31]

Thus, about 50,000 years ago, we find an explosion of creativity among the Cro-Magnons, indicating art, religious consciousness, and probably the presence of language. The question is: What triggered this revolution?

There is a great deal of speculation concerning this, but little hard evidence. The origins of consciousness, and even its definition, is one of the most disputed questions in psychology. Generally, consciousness is said to involve self-reflexive insight, and to depend not only on a large brain with a developed neocortex but also on complex social interactions, for without these interactions, consciousness does not develop fully. Understood thus, consciousness is usually explained as an emergent quality of the complex brain and the social interactions of the person. Emergent qualities are those that only emerge as properties of complex wholes. They can be predicated of the whole, but not of the parts that make up the whole. A standard example is water. When its components, hydrogen and oxygen, unite into a molecule of water, genuinely new properties emerge. So also with common salt, sodium chloride, which exhibits properties that are not found in its parts, sodium and chlorine. The argument then is that consciousness is a property that is a property of the whole brain, but not of the parts of the brain. This is how Tattersall explains it. Human beings, he says, are parts of nature, produced by the same processes that produced other natural beings. But we alone can reflect on ourselves and our difference from the rest of the natural world.[32]

From this it can be seen that the evolution of humanity is a very gradual process, with little change until the emergence of language, art, and symbolic thinking, probably some 50,000–60,000 years ago. Up until that time, there seem to be "prehumans," who were more like apes who have learned how to make some rudimentary stone tools than they were like modern humans. In the scientific community, this whole process is explained naturalistically. It is a result of purely natural processes, requiring nothing beyond nature, nothing nonmaterial. There is no place for a soul in the modern evolutionary account. I will suggest below, however, that the evidence can be read in such a way that it is compatible with the creation of a human soul, which is the source of the religious consciousness exhibited by early humans (in the Cro-Magnon burial rites) and all modern cultures.

■ Neuroscience

The study of the brain has developed rapidly in the last few decades. It has been aided by the development of new technologies, such as the

CAT scan, MRI, and PET scan, which allow investigators to create an image of what parts of the brain are active during various kinds of mental activity. This in turn has helped neuroscientists locate particular mental functions in particular brain areas. For example, the ability to recognize faces seems to be located in one part of the brain. Occasionally, persons who suffer strokes in that area suddenly lose their ability to recognize faces, even of persons they have known for years. Investigators using CAT scans have been able to locate the damaged areas of the brain.[33] Some functions associated with language, however, seem to spread across more than one region of the brain, and to involve dispersed neural networks. Brain damage can dramatically change emotional, social, and moral behavior. The famous case of Phineas Gage is often cited. Gage was a railroad foreman when, in 1848, an accidental exploding charge sent a tamping iron through the front part of his brain, entering his left cheek and exiting through the top of his head. Amazingly, he remained conscious. But afterwards his personality changed. He had been efficient and capable, but became "feckless and irresponsible, his likes and dislikes, his aspirations, his ethics and morals were altered."[34] Similarly Alzheimer's disease, which involves brain deterioration, can have profound effects on patients' personalities.

For reasons such as these, most neuroscientists believe that mental events are directly explainable by brain processes. They deny the existence of an immaterial soul, and think that sooner or later all processes of the mind—thought, emotions, and even morals and religion—will be explainable by brain processes. Neuroscientist Michael Arbib writes as follows:

> Mind has properties (self-consciousness, wonder, emotion, reason) that make it *seem* more than "merely material." . . . Nonetheless, I believe that all of this can be explained in terms of the physical processes of the brain. . . . We have already seen that, although the case is not closed, there is a respectable case for an approach to neuroscience predicated on the view that all of this can be explained *eventually* by the physical processes of the brain.[35]

Of course, we have known since the discovery of alcohol that physical and chemical processes can affect our thinking. What is new is the enormous expansion of this knowledge.

Thus both the development of the theory of human evolution and the findings of neuroscience point toward the conclusion that the human person is a psychosomatic unity, and not a soul inhabiting a body, as Descartes thought. This conclusion is by now virtually universal among scientists. Neuroscientists John Eccles and Wilder Penfield both defended a mind-body dualism, but both are now deceased, and no prominent neuroscientist has followed them in this position.[36]

Given, then, the psychosomatic unity of the person, what positions follow from that? Broadly speaking there are two: reductionistic naturalism and emergentism. I will discuss these in that order.

■ Reductionistic Naturalism

Reductionistic naturalists believe that the person can be explained by the action of her parts. The method of many sciences, such as physics, chemistry, and biology, is to explain any whole as due to the interaction of its parts. How do we explain the nature of water? By breaking it down into hydrogen and oxygen, examining the structure of its molecule, and computing how it will react to various forces. In this way we can understand the behavior of water, but also why it behaves as it does. The miniscus effect, for instance, is due to the fact that the hydrogen atoms of the water molecule form weak bonds with each other, thus giving water a film, which clings to the sides of glasses.

Most of physics and chemistry depend on reductionism for their explanations. Chemists explain compounds by analyzing the composition and structure of their molecules. Matter has been explained by the atomic theory, which explains matter by reducing it to the atomic elements. Physicists explain atoms by reducing them to their constituents: protons, neutrons, and electrons. These are further reduced to their constituents: quarks and gluons.

Reductionism has returned huge dividends to the scientific community. The whole development of atomic energy and synthetic chemistry flows from the reductionistic method in science. Molecular biology is also reaping huge rewards from the use of the reductionistic method. The explanation of inheritance by genes and DNA is one result. The ability to craft synthetic drugs is another.

The question is: Can reductionism explain the human person? Predictably, many scientists and philosophers say "Yes." The human person is the product of his or her genes, chemistry, and physics. There is no soul or vital force or anything in the person that cannot be explained and analyzed by science.

A parallel to explaining the human person reductionistically is the explanation of life. For millennia, it was held that living beings were the result of a special soul. This is the typical viewpoint of all animistic religions, which tend to endow everything, even rocks, with souls. But Aristotle also argued that living things were characterized by a kind of soul, the "form," an inner dynamic principle that made them what they are. And even into the early twentieth century, philosophers such as Henri Bergson argued for the existence of a life force, an *élan vital*, which animated living things and made them different from nonliving things. Yet virtually no biologist today accepts "vitalism" as an explanation of life. Living things, biologist Francis Crick explains, are simply highly organized complex systems, capable of utilizing energy, reproducing, and passing along the information that encodes their structure to the next generation. They are no different in kind from machines, and are often described in mechanical language by biologists.[37] Living things, then, have been explained reductionistically, by the operation of their parts.

It is only another step to apply this same analytic procedure to human beings. Crick has done this himself in a recent book, *The Astonishing Hypothesis*. There he writes: "The Astonishing Hypothesis is that 'You', your joys and your sorrows, your memories and your ambitions, your sense of personal identity and free will, are in fact no more than the behavior of a vast assembly of nerve cells and their associated molecules."[38] Crick's special project is to explain human (and animal) consciousness as the product of neural networks. There is, in effect, no "I" behind the eyes of the person, only neural networks. There is no free will; this is an illusion.[39] And of course there is no soul. Crick is contemptuous of philosophy and especially of religion; the only satisfactory method of explanation is natural science. If we want an accurate account of consciousness, therefore, we must eschew introspection, philosophy, and certainly religion, and begin with the elements of the brain—neurons—and how they are organized into neural networks. Once we understand this, we will be able to understand consciousness itself. "The aim of science is to explain *all* aspects of the behavior of our brains, including those of musicians, mystics, and mathematicians."[40]

A corollary of Crick's position is that we should be able, in theory at least, to construct thinking machines that are also conscious. For if those machines were similar to the human brain, there is no reason they should not be conscious.

Reductionist explanations are also the leitmotif of sociobiology. Sociobiologists such as Edward Wilson and Richard Dawkins aim to explain not only human nature but also human *behavior* by reduction to molecular biology. Human behavior, which was once the province of psychology and sociology, will be explained by biology—specifically by genetics. Biology in turn can be explained by chemistry, and chemistry by physics. Thus Richard Dawkins, in *The Selfish Gene* and other books, argues that human beings are, in effect, the genes' way of making other genes. Much of my behavior is simply due to my genes attempting to maximize their survivability. Dawkins writes: "We are survival machines—robot vehicles blindly programmed to preserve the selfish molecules known as genes."[41]

The recent trend in sociobiology is to explain human moral behavior in terms of genetic influence. Altruistic behavior, Dawkins avers, is really genetic selfishness. If I sacrifice my life for my brother or child, it may actually increase the chance of survival for my genes. Why? Because my brother or child carries 50 percent of my genes. If they survive and have progeny, then the genes will have maximized their chance for survival.

Religion is explained by sociobiology along similar lines. Religion intensifies human bonding, and so cements loyalties within tribes. This increases the chance of survival for individuals, and hence for their genes. Sociobiologists have been strongly criticized for ignoring the effects of culture on human behavior. In response, some of their claims have been tempered. But, as Wilson puts it, "genes hold culture on a leash."[42] It is humans who created culture, and human behavior is strongly influenced by genetics.

■ Emergentism

Emergentists agree with reductionists that the action of the parts of the human affects the whole. Human emotions and behavior, for instance, can be drastically altered by drugs, or by changes to the brain. They agree that human beings come to be through a gradual process of evolution, that there is no immaterial soul, and therefore that humans only differ from animals by degree. Much of the biochemistry of humans and primates is similar; much of the DNA is similar. But emergentists hold that unique properties emerge at the level of the whole, properties that are not predicable of the parts. Ian Barbour puts it this way:

> I take *emergence* to be the claim that in evolutionary history and in the development of the individual organism, there occur forms of order and levels of activity that are genuinely new and qualitatively different. A stronger version of emergence is the thesis that events at the higher levels are not determined by events at lower levels and are themselves causally effective.[43]

We have already considered water and sodium chloride as examples of emergent properties. Let us consider some organic ones. In the same way that a sentence can carry information, but the individual letters cannot, a molecule of DNA can carry coded information that is necessary in the organization of the cell. The cell itself has properties, such as the ability to regulate its interior environment, seek food, and reproduce, that are not found in the parts. And higher organisms carry additional properties besides: growth, locomotion, sensing, and so forth. None of these properties is found at the level of the parts, only at the level of the whole.

Within the human person, as Barbour has pointed out, there are many levels of organization (and of function).

> A living organism is a many-leveled hierarchy of systems and subsystems: particle, atom, molecule, macromolecule, organelle, cell, organ, organism, and ecosystem. The brain is hierarchically organized: molecule, neuron, neural network, and brain, which is in turn part of the body and its wider environment.[44]

At each of these levels, new properties emerge that are characteristics of the whole but that cannot be predicated of the parts.

On this model, then, consciousness is an emergent property of the whole person. It is centered in the brain but emerges as the brain/person develops and interacts with the social and physical environment. The important point is that consciousness cannot be simply reduced to the brain or to its parts. Individual neurons are not conscious, but the whole brain/person is. Though consciousness depends on the brain (in its environment), it cannot be reduced to the brain. In philosophical language, mental events are said to *supervene* on physical events but are not identical with them. That is, they depend on the physical events of the brain but may transcend those events (depending on one's definition of supervenience). When I am writing this sentence, presumably my mental activity correlates with some neuronal activity in my brain. But the thought expressed in the sentence could not be explained by merely referring to those neuronal events. Even if those events could be described precisely, one could not predict from them what thought or sentence I was writing. Another way of expressing this same idea is to say that we need two separate languages to describe what the brain does: third person language, which describes the neurological activity as an object, and first person language, which is the person's own subjective account of her conscious activities. This approach is followed by psychologist Malcolm Jeeves and others.[45] Jeeves notes: "the cumulative trend of neuropsychological research pointing to the conclusion that neural and mental processes are two aspects of one set of physical events."[46]

There are, however, different meanings of emergentism. Francis Crick, as we have seen, does affirm the existence of emergent properties, but he thinks that these are no more than the sum of the properties of the parts. That is, they can be predicted and explained by reduction to the parts. Furthermore, he would deny that emergent properties can effect what is called "downward causation," i.e., that they can cause changes in the neurons of the brain, and so initiate actions. For Crick, all the causation takes place at the level of the neurons. It is merely an illusion to think that my putatively free decision causes a change in the neuronal states. Rather, it is the neuronal networks that cause my decision, which is therefore not free. But Barbour, Murphy, Brown, Philip Clayton, and many others claim that emergent properties, especially consciousness, can effect causal changes in the neuronal networks of the brain, and can therefore initiate free decisions. Nobel laureate Roger Sperry expresses this well.

> As a brain scientist, I now believe in the causal reality of conscious mental powers as emergent properties of brain activity and consider subjective belief to be a potent cognitive force which, above any other, shapes the course of human affairs and events in the world.[47]

In this discussion of emergentism, I am assuming this second definition: that emergent qualities can affect the physical processes that engender them.

Again, virtually all those involved in the dialogue between science and theology, and many scientists, favor an emergentist view of the person. It is probably fair to say that this is the consensus position among scientists who reject the hard reductionism and agnostic materialism of Crick et al. Many of these people are Christians. But they reject what they call Cartesian dualism: the belief that there is an immaterial soul which functions independently of the body. So, if they are to preserve essential Christian doctrines like freedom of the will and afterlife, they are left with an emergentist account of the person. This account is compatible with the findings of biology and neuroscience, but, it is claimed, also can preserve what is most preciously human: the subjective aspect of our consciousness, our freedom, our sense of moral responsibility, our sense of being in relation with God.

Since emergentists reject an immortal soul, their only hope for an afterlife is a belief in bodily resurrection. For if the person is a psychosomatic unity, and not a body-soul composite, bodily resurrection would seem to be the only possibility for an afterlife. And many of those who espouse an emergentist position, such as John Polkinghorne, do affirm the reality of bodily resurrection. Others, however, such as Arthur Peacocke and Ian Barbour, do not explicitly affirm a bodily resurrection. They may affirm the existence of an afterlife, but, given their rejection of an immortal soul, it is not clear how this would be possible without a resurrection.

■ Criticism of Reductionistic Naturalism

Can we really think of ourselves as nothing but neural networks that respond in a deterministic fashion to whatever stimuli are presented to us? Crick might say we do not think of ourselves in this way, but we should learn to do so. "The Astonishing Hypothesis states that *all* aspects of the brain's behavior are due to the activities of neurons."[48] It is mere folk psychology that there is a "homunculus" (Latin: "little man") in us which decides freely between one response or another. "I" do not decide; it is the neuronal networks in my brain that react, as they have been programmed to do.

There are a number of problems with this position. First of all, while we can deny that free will exists, we cannot live as if it did not. What is called in philosophy free will is really freedom to choose and to act otherwise than we do. It does not mean simply freedom from external constraint, but also freedom from internal constraint. Now, we cannot actually live without making decisions. And in making decisions we do not act as if we are merely reacting in a predetermined, programmed fashion, as a computer would. This applies to science as well. If what Crick writes in *The Astonishing Hypothesis* is true, it means his thoughts are determined to be what they are by the neural networks of his brain. And my reaction, whether of agreement or disagreement, is likewise determined to be what it is. I am not really free to change my mind. So it is hard to see why I should seriously consider his argument. He cannot argue otherwise than he does, nor can I respond otherwise than I do. If Crick is right, then all of our decisions are on the same level as the programmed decisions of a computer. Yet in the very act of deliberating about his argument, I am assuming that I have the freedom to choose what is true or reject what is false. So it would seem that the belief in free will is implied in the very act of deliberation, whether I deny free will or not. This would seem to be true, and to be proven to be so, every time I make a rational decision.

This has implications for natural science itself. If determinism becomes the reigning philosophy in natural science, it will doom science to subjectivity, since people will always be able to argue, "That is only true for you, since your brain is wired that way, but it is not true for me, since my brain is wired differently." All claims to objectivity will be lost, for it will be impossible for anyone to rise above his or her perspective. Ironically, this is what many postmodernists claim. Everyone's perspective is socially and culturally conditioned, so no one can be objective; therefore one perspective or one truth claim is as good as another. But such a mind-set will undercut the objectivity and authority of science.

Jaegwon Kim, a philosopher, argues a similar point from the perspective of philosophy of mind. His question is: Can one idea cause or lead to another? The whole process of argument, in philosophy, law, and natural science, presumes that one idea leads to another, and so causes it. This is known in philosophy as the doctrine of mental causation. But this means that there must be room in the mind for ideas to cause other ideas. Yet if every idea is correlated with a particular state of a neural network, and that state is caused by a previous state of the same network, it is hard to see how ideas can cause other ideas. Let us assume that state N1 of a neural network corresponds to idea I1, and further that state N1 of the network causes state N2, which corresponds to idea I2. The thinker claims that idea I1 led to idea I2. But in reality, the entire causation lies at the level of the neurons: N1 causes N2 (this is precisely what Crick claims). It was not idea I1 that led to idea I2, but neuronal state N1 that determined neuronal state N2. So the belief that idea I1 led to idea I2 is illusory. There is in fact no mental causality. But this destroys reasoning, and with it science.

Of course, it is true that the brain is not a closed system, but is all the time receiving input from the external environment. But this fact does not really change the above argument, for the responses to the external stimuli are also programmed and determined (as in a computer). But what about learning? After all, we do learn to avoid mistakes, both in behavior and reasoning. Computers can learn, too, if they have the appropriate feedback. And animals also learn. They learn, for example, how to find food. Those that do not learn are weeded out by natural selection. So Crick might respond that natural selection has shaped our brains and neural networks so that even though all the causality is at the level of the neurons, nevertheless we learn to reason correctly, because if we reasoned incorrectly, we would be eliminated by natural selection. But this makes mere survival the test of correct thinking! And even if this were true, it is hard to see how creative thoughts could arise, or how we could have new insights, leading to new unprogrammed chains of reasoning, if there is no mental causation. And it is even harder to see how we could be capable of high levels of abstract thought, such as are found in theology or quantum physics, if our reasoning powers are shaped up by natural selection to simply give us an advantage in physical survival. This point was made by Alfred Russell Wallace, who was, with Darwin, the codiscoverer of the modern theory of evolution. Wallace thought that natural selection, by itself, might endow us with a brain a little better than that of an ape, whereas in fact we have a brain that is enormously superior. Recently John Polkinghorne has said much the same thing: "I can't believe . . . Einstein's ability to invent the general theory of relativity is a sort of spin-off from our ancestors' having to dodge saber-toothed tigers."[49]

The eclipse of free will also entails the elimination of moral responsibility, because no person has the freedom to do other than she in fact does. Thus those who have sexually abused children in their care are not culpable, for they could not have done otherwise. We should perhaps incarcerate them, just as we enclose dangerous animals, but we should not impute moral responsibility or guilt to them. Alcoholics who kill people on the road or who fail at their jobs are not responsible for their failure; they could not do otherwise. They have a right to treatment, and should not be punished for their failures. And so on. In such a world we would be obliged to treat other people (and ourselves!) in the same way that we treat animals, whom we also do not hold responsible for their actions. Politically, such a vision is not compatible with freedom and democracy, which is predicated on the ability of people to make responsible moral decisions. If people do not have free will or free powers of reasoning, then they should be treated politically like animals, coerced or trained into the correct behavior. But this implies that the appropriate form of government would be a benevolent totalitarian government.

Finally, it is not clear that naturalism can explain the problem of qualia. Qualia are the subjective, experienced contents of consciousness. While naturalistic accounts, such as Crick's, concentrate on the mechanisms of the brain's neuronal networks, our experience of ourselves is entirely different. We do not experience ourselves as a collection of neuron networks. A simple example may make this clear. Virtually everyone is able to picture things in their minds, say a familiar face, a landscape, or a simple object, like a red bowling ball. But there is no screen in the brain on which such an image appears. The image, as image, does not exist in the brain at all. There is no light in the brain, first of all, and second, there is no screen on which the image appears. It is perfectly plausible that the data encoding for an image of a red bowling ball might exist in the brain, just as the data for specific images exist in a computer. But the computer cannot produce the actual image without projecting it on a screen. So how is it that I, or you, are able to imagine visual images in our minds? What we experience and the physical network that encodes it in the brain seem to be entirely different, not just by degree but in kind.

Crick is aware of this problem, and he tries to address it. The screen, he thinks, is formed by the neuron nets themselves. "Suppose the screen were made of an ordered array of nerve-cells. Each nerve cell would handle the activity at one particular 'point' in the picture."[50] Apparently this means that when I remember an image, thousands of neurons are activated in a neuron network in my mind, so that each neuron is like a pixel on a screen, and together they all make up the image. But it is still not clear why I experience it as an image. For to experience it as an image, the "I" that perceives the image must be separate from the image itself, as the television viewer is separate from the screen. But such a separate "I" is precisely what Crick denies. Furthermore, if a brain scientist were able to open my brain and look at the neuron network, would he see an image? No, what he would see is a neuron network, like a network of electrical circuits, without colors, the illusion of depth, and (when it is enclosed within my brain) without light. So how is it that I experience an image, when (according to Crick) there is no "I" to see the image, and the "image" itself—the pattern of neurons—is dark?

Philosopher David Chalmers has called this the "hard problem" of consciousness. The easy problems of consciousness concern the mechanisms of cognition. These can probably be solved by neuroscience. But the hard problem is different: "The hard problem . . . is the question of how physical processes in the brain give rise to subjective experience. . . . [N]obody knows why these physical processes are accompanied by conscious experience at all."[51] To illustrate this, he gives a thought experiment. Suppose that Mary, a neuroscientist, knows everything about the brain processes responsible for color. But Mary has lived in a black and white room, and has never experienced color. She knows all about the physical and neural processes responsible for color, but she has never had the subjective experience of color. "It follows," writes Chalmers, "that there are facts about conscious experience that cannot be deduced from physical facts about the functioning of the brain."[52] Reductionist naturalism, therefore, cannot explain subjective, conscious experience. Its tendency is to reduce it to physical structures and processes in the brain. But subjective experience is what is most real and most immediate to each one of us. It is what is most distinctively human.

Naturalism, then, whether scientific or philosophical, leads to conclusions about the human person and society that are repellent, and which most persons, including many naturalists, would probably want to disclaim. In Francis Fukuyama's words, naturalism seems to lead to a "post-human" future.[53] The very core of the human person, free choice and the free power of reasoning, is annihilated. Humans are reduced to molecular machines or clever animals.

It is just these sorts of conclusions that the emergentists are trying to avoid by affirming the existence of free will and mental causation while still being faithful to the findings of modern neuroscience. And yet they, like the reductive naturalists, deny one of the oldest intuitions of the human race, namely the belief that there is a component of the human person that survives bodily death. Emergentism seems to have become the consensus philosophy of those seriously involved in reconciling Christian theology with contemporary science. Are their claims consistent? I will argue that they are not, and propose an alternative theory of human nature in the following chapter.

7

Human Nature
Embodied Self and Transcendent Soul

◼ Emergentism Again

Emergentism is the theory that new properties emerge as physical systems become more complex.[1] Emergentism differs from reductionism because it affirms that the whole is greater than the parts and so cannot be entirely explained on the basis of the parts. It is therefore a form of holism.

Emergentism is relevant to the theology of the human person because it has become, among scientists, the only real alternative to hard reductionism, and, among those trying to reconcile science and Christian faith, the favored way of explaining the human person, especially the relation of the mind to the brain. Mental properties such as consciousness,[2] self-awareness, and intentionality are said to emerge as the brain becomes more complex and interacts with its environment. They are not due to any supernatural action on the part of God or any spiritual principle, like a soul or spirit.

Emergentism is closely related to what is called "nonreductive physicalism." This theory holds that the human being is a unitary physical entity without any separable nonphysical principle (a soul or spirit), but that the person cannot be simply reduced to her physical components (as, for example, Francis Crick, a *reductive* physicalist, would hold). Nonreductive physicalists, such as Nancey Murphy, psychologists Warren Brown and Malcolm Jeeves, and others believe that, while mental events depend on physical events, they go beyond the physical events. Malcolm Jeeves puts it this way: ". . . could it not be that the same material 'stuff'—brains of animals as well as humans—due to changes in structural complexity, at some point undergo something analogous to a 'phase change' so that new properties of mind, consciousness, and a capacity for spiritual awareness emerge in humans?"[3] Most emergentists are also physicalists, that is, they believe that the person is a physical entity with no nonphysical principle present. A few, however, like Philip Clayton, resists this label.[4] My comments and criticisms below will be addressed primarily to "physicalist emergentism," that is, to that form of emergentism that assumes physicalism.

Emergentism accords nicely with the scientific findings that higher animals seem to have degrees of consciousness and mental abilities. Human beings, then, would, in the emergentist view, differ from animals in degree but not in kind. Humans have more developed neocortexes than the higher apes, and so are capable of greater degrees of abstract thought, symbolic language, and so on. Furthermore, emergentism accords with the increasing recognition that brain states affect mind states and that the human self is relational and develops through social interaction with others, not simply in isolation. So, in all these ways, emergentism fits with the scientific picture of human nature that has developed out of evolutionary biology, psychology, and the social sciences.

I agree with emergentism in many ways. It is certainly true that new properties emerge as entities become more complex. Virtually everything we see around us manifests emergent properties. Water, for example, has very different properties from its components, hydrogen and oxygen. Emergentism is a good description of this fact. It may account for many of the properties of complex organisms, like consciousness in higher animals such as dogs, primates, etc. The question is, however, does it explain the aetiology of these properties, or merely describe the process? Harold Morowitz notes that selective or pruning principles are critical for emergentism. These principles select which forms, out of an almost infinite range of possibilities, actually become real in nature. And these selection principles are largely unexplained.[5] I also agree that human beings in this life are embodied and social selves—that is, our selves are affected by our embodiment (our genes, for instance) and the society within which we mature. There is no doubt that the physical constitution of our brains and neural networks affect our minds, or that our selves develop through social interaction with others. Thus nothing I have to say in the following chapter should contradict the findings of neuroscience or the social sciences, though I may contradict some of their assumptions. Furthermore, I agree with emergentists that the whole affects the parts—indeed, I argue below for holistic causality. I concur then with Philip Clayton, Nancey Murphy, and others that our thoughts can exercise "downward causation" and initiate changes in the neural systems of our brains and even forge new neural pathways.[6] Where, then, do I disagree? My main disagreement is that it follows from the emergentist position that the self dies with the body. (No emergentist that I know of holds that there is a self or soul that survives after death.) Emergentists might (or might not) believe that individual human beings are resurrected through Christ at the end of history, but would have to explain this as a re-creation of the person, for in their view there is no self or personal identity which continues from the deceased to the resurrected person. This differs from the traditional Christian explanation that persons survive death as souls, which at the end of time take on resurrected bodies.

I also question four other points of emergentist theory. (1) Can it explain how human beings can have a personal relationship with God? (2) Can it explain the presence of genuine human freedom, manifested in free choice and free reasoning processes? (3) Can it explain how *God* can have conscious properties, such as knowledge, awareness, intentionality, and be capable of personal relationships? (4) Can it explain the evidence of near-death experiences? I believe that emergentism is inadequate to explain each of these features of human life. It is thus an attractive but incomplete theory. Therefore I will sketch out an alternative theory of human nature, which not only coheres with the findings of biology, neuroscience, and the social sciences but also explains the concerns stated above.

Before addressing these concerns, however, it is important to consider what's at stake here. If emergentists deny the existence of a soul but (some of them anyway) affirm afterlife through the resurrection, what's the problem? Let me clarify why I think a lot is at stake here.

First of all, the resurrection without any soul, that is, without any continuing personal identity of the deceased after death, is a hard sell. Most people in the pews believe that the dead are somehow still alive, right now, in a better place, with God. They are much less certain about resurrection, which flies in the face of all we know about reality (after all, we don't see resurrections occurring these days), and which seems like a vague promise a long, long way off in time. When will the general resurrection occur? Billions of years from now? Preachers might try to convince their flocks that, after all, the dead are in eternity and not in our time, and so are "already" resurrected. But given that the mourners can see the body of the recently deceased before them at the funeral, this also will be a hard sell. After all, Jesus' body disappeared from the tomb, according to the Gospel accounts, indicating that it was the

155

same body as the one that died, and that there was some connection with that body in the present and its resurrected form in the afterlife. That does not happen to people today—their bodies do not disappear from the graves. So, if Jesus' resurrection is a promise and model of our own, it follows that the dead are still awaiting resurrection, which is exactly what traditional Christian teaching says. Second, many people (Catholics as well as others) believe in prayer for the dead and in some form of Purgatory. It is traditional Catholic doctrine that the saints in heaven are not yet resurrected; they, like us, await the resurrection. Furthermore, it is questionable whether they really are fully in eternity, since they interact with us in time. All such beliefs—that the dead are even now with God, that we can pray with and to the saints, that there is a kind of purificatory process after death—are likely to disappear if the emergentist theory of the human person triumphs.

Because of this, while some intellectuals may be able to believe in a resurrected afterlife but not in souls, I doubt that the majority of believers will be convinced by this. If so, the consequence will be that in Christianity belief in the afterlife will fade away, like the grin on the Cheshire cat. This has already occurred in Reform Judaism, and it seems to be happening now in liberal Protestantism. But such an effect would be disastrous for Christianity precisely because so much of the traditional religion depends on belief in an afterlife. After all, Christianity has never taught that the righteous, the poor, the oppressed, and Christians themselves find perfect union with God or complete fulfillment in this life. They find it in the next life. The Gospels are full of sayings like this: "Do not store up for yourselves treasures on earth, where moth and rust consume and thieves break in and steal, but store up for yourselves treasures in heaven, where neither moth nor rust consumes and where thieves do no break in and steal" (Matt. 6:19–20). Without an afterlife we would be better off to hoard our treasures on earth, build bigger barns (cf. Luke 12:13–21), and leave the poor and oppressed to their fate. The rich, Jesus says, have their reward now, but if now is all there is, many Christians will follow their example. I believe that the greatest reward Christians can experience is the presence of God in their lives. But I also have learned that this life is full of hardship, struggles, trials, disappointment, and persecution, and that the fullness of God's presence does not come until the afterlife. A loss of belief in afterlife would eviscerate Christianity and the Christian practice of everyday life. Support for my thesis that an emergentist view of human nature would result in a loss of belief in an afterlife can be found in the dialogue of theology and science itself, where emergentism is the favored model of human nature, and serious discussion of afterlife is seldom broached.

Thus the credibility of an afterlife is my main concern with emergentism. But I also doubt that emergentism can support a robust defense of human freedom, our experience of God's presence in this life, or even that God has knowledge, awareness, consciousness, and mind. And loss of these would also be crippling for the Christian practice of everyday life. To these points I now turn.

■ Personal Relationship with God

We all know that it is easy for people to imagine gods or a God, a mighty Being in the sky who can protect and comfort in times of affliction. But the experience of deeply religious theistic believers is not simply imagining and placating God as some external object. Authentically religious people experience a personal relationship with God. They experience God's call on their lives, God's response to their prayers, and God's presence in their lives. As the Psalmist puts it: "You show me the path of life, in your presence there is fullness of joy . . ." (Ps. 16:11). The Hebrew and Christian Scriptures, as well as the Scriptures of other religions, are replete with accounts of such experiences. Saul of Tarsus, for example, had an intense experience of the risen Jesus, as described in Acts 9—an experience so intense that it transformed his entire life. In later writings he describes his experience of personal closeness with the Lord Jesus: ". . . it is no longer I who live, but it is Christ who lives in me." (Gal. 2:20). Mystical experiences (which are found in all religions) are particularly dramatic and intense examples of the personal experience of the presence of God. But even ordinary believers who are not mystics also experience God's presence in their lives, though perhaps to a lesser degree. Atheists, agnostics, and persons who study religious phenomena from the outside are probably not aware of how intense and important the sensed presence of God can be to believers. Hence they typically try to explain religion in terms of extrinsic effects. For example, it assuages the fear of death, it provides believers with a heavenly father-figure, it promotes group bonding, etc. But as one who has tasted the presence of God, I can assure them that this is not the core of religious experience at all. The core is the personal experience of the living and loving God in one's life.

Now the question is: If this experience is genuine, that is, if it really is an experience of an Other and not just of some aspect of the self, how can it be explained by an emergentist model of human nature? We can explain how it is possible to have relationships with external objects, including persons, because they can be perceived by the senses. But how is it possible to have a relationship with an immaterial Other who cannot be known by the senses? I cannot see how an emergentist model can explain this. I think emergentists would have to say that a personal relationship with God was not really a relationship at all, but merely a subjective experience, an experience of an idea or of part of one's self ultimately explainable by neural networks in the brain. But at least for those who have tasted the presence of the living God in their lives, this is an inadequate explanation.

▉ Freedom

Freedom, as we have seen, is essential if human reason is to be capable of arriving at conclusions that are not determined in the sense that the calculations of a computer are determined by its programming. Freedom in this sense is not freedom from external constraint, but freedom from internal constraint. A computer may be free from external constraint—no one forces it to come to a specified conclusion—but it is not free from internal constraint. Given the same input, it will always follow the same pathway. Human free choice means that we can choose to do otherwise than we in fact do (whereas a computer cannot choose to do other than it is programmed to do). We can resist strong inclinations and desires, such as anger, sexual urges, and other cravings, and we can do so not just from fear of punishment, as a dog might do, but because of moral convictions. We might have reasons to do something, but the reasons do not compel us.

Now the question is: How can authentic freedom develop at all in a wholly material system, however complex it is? The emergentist argument is, roughly, that as the brain becomes more evolved, with a neocortex and a capacity for self-awareness, freedom develops also. But it is hard to see how this can be. Self-awareness may be a necessary condition for free choice, but it is not a sufficient condition. Material particles are not free. Nor are vast collections of them (like stars) free, nor complex systems of them, like computers, free. Quantum particles, it is true, exhibit indeterminacy. But indeterminacy is not the same as freedom. Free choice has an element of intentionality about it that is lacking in quantum indeterminacy. The state taken by a quantum particle when the wave packet collapses is statistically random, whereas the result of a free choice, even an evil choice, is not random but intentional, and it exhibits purpose. Now, many thinkers, for example Philip Clayton or Nancey Murphy, who identify with the emergentist position, also defend free will. My question is: How does it cohere with their systems? Material systems, even chaotic systems or highly complex systems (like computers) might exhibit randomness (and therefore unpredictability), but they do not exhibit freedom. Genuine freedom, I maintain, comes about only because God establishes a personal relationship with human persons at a very early stage in their development.

■ Can God Know?

It is essential to the biblical vision of God that God is personal, knows his creatures, has personal relationships with individual human beings, loves them, responds to prayers, and therefore has awareness, knowledge, and intentions. The question is: How is this possible? According to an emergentist model, the ability to know, be aware, and to have intentions all depend entirely on a complex material brain. If the brain dies, there is no consciousness. So how can God be conscious, aware, intentional, or knowing since God is not material?[7] On the emergentist model, we would have to say that God's consciousness and knowledge would also have to be dependent on a complex material matrix.

Now, before answering this, we have to acknowledge that God's knowledge is not like our knowledge. First of all, God knows all that can be known, and knows it all at once, whereas we know sequentially by forming arguments, thinking things through, etc. And our knowledge is limited, whereas that of God is unlimited. So God's knowledge differs from ours as the infinite from the finite or the perfect from the imperfect. But, nonetheless, if the biblical images of God knowing and loving mean anything, there must be an analogical similarity between our knowledge and God's. If there is no similarity, if God's consciousness, knowledge, and awareness and ours are entirely dissimilar, then we cannot affirm anything about God, even that God is personal or that God knows us as persons or answers prayers. And this would mean that biblical revelations about God were meaningless. Therefore, following the principle of the analogy of being, I hold that there is a core similarity between what we mean by knowledge, awareness, and responsiveness, and what those mean in God. But if this is true, how can emergentism explain God's consciousness?

According to emergentism, knowledge is only possible because we have highly complex physical brains. As our brains develop, we can know more. We know more than spiders, worms, dogs, or monkeys because we have more developed nervous systems and brains than they. But if knowing depends on a complex nervous system and brain, how can God know? God does not have a complex nervous system, or a brain; indeed, God is not physical at all. So, on the emergentist account of knowing, God should not be able to know anything at all. Now, perhaps we could say that God's body is the whole universe. But this does not solve the problem. The universe is not complex in the way the human brain and nervous system is. Most of the universe is empty space. Even stars and galaxies, though vast, are not complex. Complexity means having many interconnections. Each neuron in the brain has perhaps 6000 interconnections with other neurons. That's complex! But stars are largely formed of hydrogen and helium with traces of other elements, and these have no stable interconnections. The interior of a star is simply a very hot gas. Furthermore, there are no interconnections between stars (except gravity) or between galaxies. There is nothing like a nervous system interconnecting the stars and galaxies of the universe. So how can God have a mind or be conscious at all? It seems, on an emergentist account, that God cannot. Or, we would have to say that God's consciousness is explainable on an entirely different principle than is human consciousness. But this leaves a marked dualism and gulf between God and humankind. And this, it seems to me, indicates a major problem with emergentism. Though it wants to avoid radical dualism, it falls into a radical and perhaps unbridgeable dualism between God and creatures. Essentially, this is due to the fact that emergentism, like naturalism, starts with the physical universe as what is most real, and then tries to work its way up to consciousness and God. But this way in the end fails. It cannot really account for free will, mental causation, or for the consciousness of God. I will suggest below that the better way is to start with consciousness or mind itself as what is most real and foundational in the universe, and work from consciousness to the material world.

■ Near-Death Experiences

My last critique of emergentism is that it cannot explain adequately the evidence of near-death experiences. These are usually ignored in discussions of the human person and the soul. This is partly because their interpretation is controversial, and partly because the evidence supporting them is largely anecdotal. But partly it is because they tend to undercut naturalism and to point in the direction of dualism, a position that is seen as intellectually indefensible by naturalists and by the scientific community generally. But I do not think that the evidence provided by near-death experiences should be ignored just because it calls into question the reigning interpretive paradigm. Rather, it is germane to the question of emergentism and to our interpretation of the human person generally.

Near-death experiences (hereafter NDEs) were made popular by psychiatrist Raymond Moody's book *Life after Life*,[8] first published in 1975. Since that time a voluminous literature has been published concerning these events.[9] They vary of course from individual to individual and culture to culture, but certain elements are found to recur in most experiences. A typical instance might occur in an operating room. The patient, who is comatose, feels herself leaving her body, traveling down a "dark tunnel," coming into a heavenlike world of great beauty, encountering a "being of light" who communicates with her through direct thought transference, seeing a review of her life, meeting other (dead) persons, being given a choice of whether to go back into her body or not, and then choosing to return or being sucked back into the body. During this time, patients experience no pain, whereas their return into their body is accompanied with great pain and distress. Usually these experiences are life changing. People become convinced that there is a life after death. They lose any fear of death. They are less concerned with material things and more concerned with spiritual things and with personal relationships, though they also may become less committed to any particular religion or denomination.

For our purposes, there are two features of these experiences that are important. First, many persons claim to be able to describe what is going on around them, or even what is happening in other places, from an out-of-body perspective, during the time when they are "dead" or comatose. This is called "remote viewing." Second, many claim to have seen dead loved ones during their experience. And sometimes those encountered were persons whose death was unknown to the patient having the near-death experience.

Elisabeth Kübler-Ross provides some striking testimony concerning both of these occurrences.

> I have studied thousands of patients all around the world who have had out-of-body or near-death experiences. . . . Many of these people were not ill prior to the life-threatening event. They had sudden unexpected heart attacks or accidents. . . . The common denominator of these out-of-body experiences is that these people were totally aware of leaving their physical body. . . . [T]hey found themselves somewhere in the vicinity of where they were originally struck down: the scene of an accident, a hospital emergency room, at home in their own bed. . . . They described the scene of the accident in minute detail, including the arrival of people who tried to rescue them from a car or tried to put out a fire, and the arrival of an ambulance. Yes, they described accurately even the number of blowtorches used to extricate their mangled bodies from the wrecked car. . . . We, naturally, checked these facts out by testing patients who had been blind with no light perception for years. To our amazement, they were able to describe the color and design of clothing and jewelry the people present wore. . . .

161

When asked how they could see, people described it with similar words: "It is like you see when you dream and you have your eyes closed."

The third event they shared was an awareness of the presence of loving beings, who always included next of kin who had preceded them in death. . . .

How does a critical and skeptical researcher find out if these perceptions are real? We started to collect data from people who were not aware of the death of a loved one, and who then later shared the presence of that person when they themselves were . . . at the "gate of no return."

She provides examples, among them the following:

"Yes, everything is all right now. Mommy and Peter are already waiting for me," one boy replied. With a content little smile, he slipped back into a coma from which he made the transition which we call death.

I was quite aware that his mother had died at the scene of the accident, but Peter had not died. He had been brought to a special burn unit in another hospital severely burnt, because the car had caught fire before he was extricated from the wreckage. Since I was only collecting data, I accepted the boy's information and determined to look in on Peter. It was not necessary, however, because as I passed the nursing station there was a call from the other hospital to inform me that Peter had died a few minutes earlier.

In all the years that I have quietly collected data from California to Sidney, Australia; from white and black children, aboriginals, Eskimos, South Americans, and Libyan youngsters, every single child who mentioned that someone was waiting for them mentioned a person who had actually preceded them in death, even if by only a few moments. And yet none of these children had been informed of the recent death of the relatives by us at any time. Coincidence? By now there is no scientist or statistician who could convince me that this occurs, as some colleagues claim, as a result of "oxygen deprivation" or for other "rational and scientific" reasons.[10]

In her early books, such as *On Death and Dying*, Kübler-Ross rejected any belief in afterlife, and considered it a way of denying the reality of death. But as the above quotations show, she changed her opinion dramatically as a result of patients she encountered (mainly children) who had had near-death experiences.

Now, how should we explain near-death experiences? Usually, they are explained by the scientific community as due to some physical cause, such as drugs, lack of oxygen in the brain, endorphins in the brain, hallucinations, or the presence of residual consciousness in the patient (the patient appears to be comatose, but is really partly conscious, and hence can perceive what is happening around her). The problem with these "explanations" is that they do not explain the evidence.[11] First, many patients were not on drugs at the time of the NDE. Second, in operating rooms blood oxygen is routinely monitored, and most patients do have enough blood oxygen. Third, endorphins do relax persons (they are the cause of the relaxation we feel after heavy exercise), but they do not cause out-of-body experiences. Further, patients on B-endorphins still experience pain, for instance, if they are stuck with a needle. But people in NDEs experience no pain, even if they are being given electroshock to revive them. Fourth, hallucinations typically vary enormously, whereas NDEs have a family resemblance. Patients themselves will deny that NDEs are hallucinations. Further, hallucinations cannot explain the remote viewing that patients claim to experience. Near-death experiences are sometimes explained by residual consciousness in the dying brain. This might explain patients' being aware of what is going on around them. But there are problems with this explanation. First, most of the claims to remote viewing are visual, not auditory, and their eyes are shut. And in many cases, patients claim to have seen things out of their visual field, which they could not have seen even if they had been fully conscious. And very often, these claims have been corroborated by others at the scene. Finally, there are many cases in which patients have had NDEs while being hooked up to EEG machines, and these registered no brain wave activity.

English psychologist Susan Blackmore has an interesting explanation, called the dying brain explanation. She explains NDEs as due to neurological effects in the cortex as the brain dies from lack of oxygen.[12] The problem with her explanation is twofold. First, in many cases of NDEs, there is no shortage of oxygen to the brain.[13] And second, if the patients' brains had been dying, one would expect there to be some form of brain damage. But there isn't. In all the accounts I have read, there is never any claim that the patients who recovered showed any trace of brain damage. Just the opposite: their recollections of their NDEs are remarkably vivid. Furthermore, the dying brain hypothesis cannot explain patients' claims to remote viewing, as Blackmore herself admits.[14]

Recently, Eugene d'Aquili and Andrew Newberg have hypothesized that NDEs are due to Jungian archetypes.[15] The problem with this explanation is that Jungian archetypes are scarcely less controversial within the psychological community than NDEs themselves. Very few psychologists today accept Jung's explanation of the psyche. Still, the explanation may prove to be of use, especially in explaining the question of how the NDEs are interpreted by the patients. But, like the dying brain hypothesis, this explanation does not explain the claim to remote viewing.

The claim of remote viewing when out-of-the body is the crucial question in trying to interpret NDEs. It is now widely accepted that patients *experience* being out of the body and seeing events from an out-of-body perspective. The question is: are they really able to do this? Are they able to perceive while being out of the body? For if they can actually perceive and describe what is going on around them, and even what is going on in remote locations, when they are comatose, with no brain waves, then that would seem to be *prima facie* evidence that consciousness and perception can survive outside of the brain. And the same is true if they are really able to perceive (disembodied) dead loved ones. If either of these claims is true, it would seem that an explanation of consciousness that insists that it is dependent on the brain, and cannot survive brain death, is false. Consciousness could exist outside the brain.

There is at least one controlled study that supports these claims, a study by Atlanta cardiologist Michael Sabom. Sabom tested whether those who claimed to have had an NDE could accurately describe the resuscitation procedure that had been used to revive them from cardiac arrests. "I was anxiously awaiting the moment when a patient would claim that he had 'seen'what had transpired in his room during his own resuscitation. Upon such an encounter, I intended to probe meticulously for details that would not ordinarily be known to nonmedical personnel. . . . I was convinced that obvious inconsistencies would appear which would reduce these purported 'visual' observations to no more than an 'educated guess' on the part of the patient."[16] After five years, Sabom had collected thirty-two such cases. But then he wondered what an "educated guess" of a resuscitation procedure would look like. So he set up a control group of twenty-five "seasoned cardiac patients" who had been "consecutively admitted to a coronary care unit and had been hospitalized from heart attacks." Thus this group of twenty-five cardiac patients had received considerable exposure to hospital routine and television programs, both of which could have contributed to their knowledge of CPR. But they had not experienced NDEs, nor had they ever actually observed a resuscitation procedure. When asked to describe such a procedure, twenty of these twenty-five patients made at least one major error; three gave limited descriptions that were without obvious error; and two claimed no knowledge of CPR technique.[17] Of the thirty-two who had claimed to experience NDEs and to have had out-of-body experiences during these NDEs, twenty-six were able to give general visual descriptions of what was happening in the room during their NDE, but were not able to give specific details about the procedure. The reason for this, they claimed, was that their attention had been occupied by the unique qualities of the experience itself, and amazement at being out of their bodies. None of these twenty-six, however, made any major errors in the description, whereas 80 percent of the control group had made major errors. Six of the thirty-two were able to give precise visual details of the procedure, which corresponded very well with the written hospital reports. These included such details such as describing the movement of the pointers on the defibrillator machine, describing the doctor's trying to insert a needle into the subclavian vein on one side of the body, failing, and then moving to the other side, and so on. One man accurately described what three of his family members were doing in the corridor during his resuscitation. The visit of these family members had been unannounced and unexpected, and they were not visible from the room in which the man was undergoing resuscitation.

Sabom concludes from this study that the apparent accuracy of visual descriptions of CPR procedures and associated events cannot be explained by "prior general knowledge, by information passed on by another individual, and by physical perception of sight and sound during semiconsciousness."[18]

In the remainder of his book he examines in detail other possible explanations, such as lack of blood oxygen, the presence of endorphins, hallucinations, excess carbon dioxide in the brain, etc. Though he had begun the study highly skeptical of the reality of NDEs , he ended by believing that the only plausible explanation was that the patients really were able to perceive the CPR procedures from a position outside the body.

> During the *autoscopic* [i.e., self-viewing] portion of the NDE, near-death survivors claimed to have seen and heard events in the vicinity of their own unconscious physical bodies from a detached elevated position. The details of these perceptions were found to be accurate in all instances where corroborating evidence was available. Moreover, there appeared to be no plausible explanation for the accuracy of these observations involving the usual physical senses. An out-of-body (extrasensory?) mechanism would explain both the personal interpretation afforded these experiences by those who had them . . . and the "visual" accuracy of the autoscopic observations.[19]

Yet the conclusion of Sabom, Kübler-Ross, and others, that persons having NDEs really are able to perceive events from an out-of-body perspective, has not gained acceptance in the scientific community. There are several reasons for this. First, most of the evidence is anecdotal; there are very few controlled studies. And, even though some sciences, such as anthropology, do depend on anecdotal evidence (such as stories gathered from field reports), most scientists, including psychologists, do not accept anecdotal evidence as reliable. We know, for example, that different witnesses of an accident will give slightly differing accounts. But if there are hundreds of anecdotal accounts, all pointing in the same direction, then it becomes irrational to simply reject them. After all, many things in life, including virtually all historical events, are known through testimonial, or anecdotal, evidence. And science itself cannot escape this, because any scientific paper, even of a controlled study, is presented to us on the basis of testimonial evidence. The strength of the evidence lies in the amount of corroborative testimony, and in the reliability of the witnesses. If many scientists have confirmed a study, it should be considered reliable. But by the same logic, if many persons, who are judged to be reliable reporters, and who have no reason for lying, give testimony to a type of event, then that occurrence should be credible.

Another reason why the remote viewing claimed in NDEs is hard to believe is that there is at present no way to explain it. It seems impossible that persons could perceive what is happening if they are not using their senses. But the fact that we can't explain how something occurs is not a reason for rejecting the evidence that it does occur. If we did that, there would be no scientific progress at all, for any discovery that was really new and inexplicable would then have to be rejected by scientists, simply on the basis that is was presently inexplicable. When Newton posited the existence of gravity as a force acting at a distance, he had no explanation for it. Indeed, there was no sufficient explanation until Einstein's theory of relativity, in the twentieth century. But, though there were many critics of Newton's idea of a force acting at a distance, eventually the scientific world accepted the existence of gravity, even though how gravity worked could not be explained for three hundred years. Likewise, contemporary physics cannot explain the so-called nonlocal reactions between entangled particles. But this has not prevented the experimental evidence from being accepted by the community.

A final reason NDEs are problematic is that the idea that consciousness can exist outside the body simply contradicts naturalism. If consciousness can indeed function outside the body, without being dependent on the brain, then naturalism (as defined in this book) would seem to be false. This is the most probable reason why Kübler-Ross's and Sabom's explanation for NDEs is rejected. Indeed, NDEs are perfectly acceptable to the scientific community if they are explained naturalistically. But if they are explained "dualistically," they are unacceptable, because dualism is unacceptable. (It is unfortunately true that one's theoretical worldview or paradigm strongly influences what evidence one considers acceptable or unacceptable.)

If we think that emergentism is not adequate as a way of accepting modern neuroscience while holding onto free will, mental causality, and the consciousness or mind of God, then we are in a dilemma. Either we must accept reductionistic naturalism, such as that of Francis Crick, which eliminates God, the soul, free will, and (apparently) mental causa-

166

tion, or we must embrace dualism. The problem with dualism, at least as described by Descartes, is that it seems to isolate the soul from the physical processes of the brain. But the work of contemporary neuroscience and psychology has shown conclusively that consciousness is dramatically affected by the physical processes of the brain. Strokes, for example, can profoundly affect a person's memories and thinking processes. Psychologist Malcolm Jeeves calls this "the tightening of the brain-mind behavior links."[20] So what do we do?

Much of this problem has occurred because the relevant literature has set up a false antithesis: either some form of naturalism and emergentism, or dualism. There are other possibilities. In the following pages, I will propose one possible hypothesis: a scheme of the human person in which the soul is the organizing principle, the holistic cause of the body. As such, the soul is affected at all points by the physical condition of the brain and body. The person is therefore, in this life, a psychophysical unity. Damage to the brain would impede the relevant mental processes. Furthermore, the embodied soul shapes its own destiny by the choices it makes in this life. Nevertheless, I will maintain that the soul and the essential features of consciousness can survive bodily death. But, I will hold, this is not a strong dualism, because there is an inner unity between matter and spirit. If matter and spirit were completely different, there could be no communication between them.

■ The Soul as Holistic Cause

Thomas Aquinas, following Aristotle, thought that the human soul was the formal cause of the human body. Formal cause, in this sense, is an organizing cause, that which causes the whole to behave as a unity. Rather than use Aquinas's terminology, however, I will use the expression "holistic cause." My proposal is that the soul is the ultimate organizing principle of the body. It is not a separate, independent substance, as Descartes thought. Rather, it is an active, internal principle that acts to keep the whole functioning as an integrated unit. We could think of it as a field of active information, which informs the whole, keeping it in order. (Parallels in physics would be gravitational and electromagnetic fields. These differ from the soul in that they are detectable by physical instruments, whereas the soul, so far, cannot be detected.) Now, as an information field, the soul is embodied, and is therefore affected by any damage to the body, especially to the brain. Imagine a software program running on a computer. If there is damage to one section of the computer's hardware, the software program will not be able to function properly. But the example of a computer is misleading in one respect. If the soul is an active organizing principle, then it would be like a software program that also affects the organization of the hardware.

How might the soul affect the physical processes of the body? The interaction might take place at the quantum level. If I make a decision, that affects the neurons of the brain. How is this done? It may be, as Nancey Murphy has suggested, that the neuron networks of the brain are affected by the decision at the quantum level, so that the neuron networks are changed, and an action results (for example, I raise my arm). This is possible because at the quantum level, the states of electrons and other subatomic particles are somewhat indeterminate. An input of information could cause a change in the state of a quantum system, and so, possibly, in the state of a neuron network. But this would have to involve many different, simultaneous, and coordinated quantum changes, not just one or two. And so the change would have to be a holistic one, not one that simply affected a point or two in the neuronal network.

Now this may seem similar to Arthur Peacocke's idea of the whole influencing the parts. But there are two differences. First, Peacocke thinks of the whole as a boundary condition that acts by constraining the action of the parts, not by the active input of information. But a holistic cause acts more directly, through the active input of information. Second, Peacocke denies that the whole acts by affecting the quantum level.

The idea that the soul is a holistic cause might also seem close to the emergentist conception of the mind and consciousness as an emergent, embodied quality. But it differs in two important ways. If the soul is a holistic cause, it is not just an emergent quality of a complex system that is already formed. It is itself the organizing principle of the whole system. Second, in traditional Christian belief, the human soul (though probably not animal souls) can survive the death of the body and brain. This is possible because God establishes a relationship with the person early in his or her existence. I will consider this point further below.

Now, is there any evidence for the existence of holistic causes, in the sense I have described them above? There are at least four areas in the sciences that suggest the existence of such causes: (1) the so-called nonlocal interactions of widely separated but "entangled" particles in physics; (2) the Pauli exclusion principle; (3) so-called "directed mutations" in biology; and (4) certain holistic phenomena in brain science, such as the findings of Wilder Penfield. I will consider these in order. (These discussions are, however, somewhat technical, and some readers may want to skip to the next section.)

168

The Pauli Exclusion Principle

The Pauli exclusion principle states that "no two electrons in an atom can have the same four quantum numbers."[21] This means that no two electrons can have the same state in an atom. Each electron must have a different spin or occupy a different shell from any other. For example, the hydrogen atom has one proton and one electron. The single electron occupies the 1s orbital in the K shell (the lowest shell). Helium has two protons and two electrons; its two electrons both occupy the 1s orbital in the K shell, but have opposite spins. However, the K shell is complete with two electrons; it can hold no more. Hence helium is an inert gas—it neither receives nor gives electrons; it can hold no more. Hence helium is an inert gas; it neither receives nor gives electrons to other atoms. The next heaviest atom is lithium, with three protons and three electrons. Its third electron must occupy a higher shell, the 2s orbital in the L shell. This electron is not paired with any other, and so can be easily transferred to, say, an atom of chlorine, which has seven electrons in its outer shell. This makes lithium a highly reactive element. Chlorine is also reactive, because its outer electron shell forms a stable whole with eight electrons, and so it tends to attract another atom with a single electron. But the reactivity of lithium and chlorine, as distinct from the interness of helium, is due to the Pauli principle, which is a fundamental law of physics immanent within matter itself.

If it were not for the Pauli principle, all electrons could occupy the same orbit or orbits, and matter would not be differentiated into elements, and further differentiated into the myriad compounds that the basic elements form. Matter would all be a "sea of electrons and nuclei at the lowest energy level."[22] All of chemistry, indeed all of complex matter and life, depends on the Pauli principle.

But this principle is mysterious. It is not related to other physical forces (like gravity, electromagnetic force, etc.) or other laws of physics.[23] And it is not clear how it works. Biophysicist Harold Morowitz writes, "Since the principle is nondynamical [i.e. it does not work through a force], it is as if the second electron knew what state the first electron was in. . . ."[24] What the Pauli principle leads to is the existence of structure and form in atoms and molecules; it is a structuring and ordering principle. And it acts at the level of the whole atom. It is not as if one electron impinges on another, forcing it into a different spin state of a different orbital. Electrons only assume specific orbitals when they are united with a whole atom, and it is only then that the exclusion principle operates. So this is a holistic ordering principle, a cause that operates on the level of the whole. Here the whole, the atom, shapes the behavior of the parts—the electrons. Morowitz concludes his discussion of the Pauli principle by writing, "Because of the Pauli principle, matter is informatic, and something akin to mind has already entered the universe."[25] The Pauli principle, then, is an example of a holistic cause or ordering principle that seems to act through the communication of information.

Nonlocal Interactions

It is possible to obtain pairs of photons (or other subatomic particles) that are in identical yet indeterminate states. For example, a single photon can be passed through a crystal of potassium niobate, which will split the original photon into two "twin" photons, whose total energy equals that of the parent photon. Now, quantum theory tells us that the state of these twin photons (their spin, polarization, energy, etc.) must be indeterminate until they interact with a measuring device. The act of measurement causes the photon to assume a definite state. Now, one would expect, since these twin photons are separated, sometimes by great distances, that each would have to be measured if each is to assume a definite state. Furthermore, one would expect that if each were measured independently, the states of the two would sometimes vary. Sometimes photon A would assume an "up" spin, and photon B would assume a "down" spin, for the state assumed by the photons at measurement is not predictable. If one measures 1,000 photons that are independent of one another, one would expect 500 to assume an "up" spin and 500 to assume a "down" spin. But in the twin or "entangled" photon experiments, if one measures one photon, the other always is found to have assumed the same state as its twin. In 1997, Nicolas Gisin and his colleagues, of the University of Geneva, performed an experiment in which they sent twin photons along optical fibers to destinations north and south of Geneva, seven miles apart. Gisin's experiment showed that when one of the photons was measured, and so assumed a definite state, the other assumed the same state in less than one ten-thousandth of the time it would have taken a light beam to travel from one of the photons to the other.[26] In other words, the second photon "knew" what state its twin had assumed, presumably instantaneously. And there was no possible communication between the photons, for, according to the theory of relativity, no signal could travel faster than the speed of light. So how do the photons "know" what each other is doing? Physics has no answer to this problem. John Polkinghorne observes:

> We are back to the idea that quantum systems exhibit an unexpected degree of togetherness. Mere spatial separation does not seem to divide them from each other. It is a particularly surprising conclusion for so reductionist a subject as physics. After all, elementary particle physics is always trying to

split things up with a view to treating them independently of one another. I do not think that we have yet succeeded in taking in fully what quantum mechanical non-locality implies about the nature of the world.[27]

There are several possible ways to explain this, but one rather obvious way is the existence of a holistic field of active information, which causes the twin photons to behave as a single entity, so that whatever happens to one happens to the other.

Another example comes from physicist John Hitchcock. Snowflakes, as we all know, are built up from atoms of water. Almost all snowflakes have a hexagonal structure, but the total possible variety of forms within this basic structure is immense. Each snowflake then has what is probably a unique pattern. Yet in each snowflake, the pattern of each arm is identical to that of every other arm. Hitchcock comments:

> A point to note carefully here is that as water molecules freeze onto the outer end of one arm of the flake, molecules freezing onto another arm somehow "know" how to freeze in precise symmetry with the first, though the distance between the two events is large compared to the size of the molecule. This could only be the case if there is an "image" being symmetrically extended as the snowflake grows, or rather not an image, but a pattern behind the highly individualized snowflake image which we see. This pattern cannot depend only upon nearby molecules. Since the crystal has the ability to correct for minor "freezing errors" and to continue with the original pattern, there must be some degree of dependence upon the whole flake.[28]

Holistic Causes in Biology

There are also some indications of holistic causality emerging in biology, namely in what are called "directed mutations," discussed in chapter 5. It is axiomatic in biology that mutations are random, and that randomness means that the likelihood of mutation is independent of its value to the organism. In other words, the organism cannot direct its own mutations. Yet Hall's experiment, and others as well, seems to indicate that, in some way not understood, organisms can, under some conditions, direct their own mutations. In a survey of this literature, Richard Lenski and John Mittler conclude as follows:

> Evidence for directed mutation has been published in major journals by several respected biologists and for many different experimental systems. Moreover, the effect is reported to be very large, and the inference that certain mutational rates occur at much higher rates specifically when they are advantageous is said to be unambiguous. . . . [T]he view that certain mutations happen much more often when they are advantageous has been

accepted by some observers. . . . But in our opinion, none of the claims for directed mutation has compellingly excluded all of the alternative hypotheses based solely on random mutation.[29]

Thus at least some biologists have accepted the evidence for directed mutation, though it remains controversial. Lenski and Mittler argue that all possible alternatives have not yet been ruled out. They also note that no mechanism explaining directed mutation has been found.

If directed mutation is indeed a fact, it would be a powerful argument for the operation of a holistic cause. For it would seem that the bacterium (or other organism) acting as a whole can influence the operation of its parts in a rather direct way.

Brain Science

A third area where we find indications of holistic causes is in brain science itself. Here the work of Wilder Penfield is of particular interest. Penfield was a Canadian neuroscientist, and one of the fathers of neurosurgery. During the 1930s, it was the practice to try to cure certain forms of epilepsy by operating on the brain. Penfield discovered by accident that if certain regions of the cerebral cortex were stimulated by a small electric current, patients would experience vivid memory flashbacks. They might hear a melody being played, recall being at a baseball game, and so on. He also found that he could temporarily inhibit the speech centers of the brain by applying the electric current to that area of the cortex which controlled speech. On one occasion he was operating on a patient he calls "C.H." The area affected by the epilepsy was close to the speech center of the brain, so in order to avoid permanently damaging the patient's speech mechanism, he decided to map out the speech area before beginning the operation. He did this by applying the electrode to various areas to determine the limits of the speech area.

> One of my associates began to show the patient a series of pictures on the other side of the sterile screen. C.H. named each picture accurately at first. Then, before the picture of a butterfly was shown to him, I applied the electrode where I supposed the speech cortex to be. He remained silent for a time. Then he snapped his fingers as though in exasperation. I withdrew the electrode and he spoke at once: "Now I can talk," he said. "Butterfly. I couldn't *get* that word 'butterfly', so I tried to get the word 'moth!'"
>
> It is clear that while the speech mechanism was temporarily blocked he could perceive the meaning of the picture of a butterfly. He made a conscious effort to "get" the corresponding word. Then, not understanding why he could not do so, he turned back for a second time to the interpretative mechanism . . . and found a second concept that he considered the clos-

172

est thing to a butterfly. He then presented that to the speech mechanism, only to draw another blank.[30]

In this interesting scenario, the patient's mind understood what a butterfly was, but could not express it because the speech mechanism was blocked. This would be similar to a person trying to raise his arm when the nerves were paralyzed, either through anesthetic or permanently.

Penfield also found that by stimulating the motor cortex he could cause the patient to move his hand or limb.

> When I have caused a patient to move his hand by applying an electrode to the motor cortex of one hemisphere, I have often asked him about it. Invariably his response was: "I didn't do that. You did." When I caused him to vocalize, he said "I didn't make that sound. You pulled it out of me."[31]

Penfield comments on this:

> The electrode can present to the patient various crude sensations. It can cause him to turn head or eyes, or to move the limbs, or to vocalize or swallow. It may recall vivid re-experience of the past. . . . But he remains aloof. He passes judgment on it all. He says "things are growing larger" but he does not move out of fear of being run over. If the electrode moves his right hand, he does not say "I wanted to move it." He may, however, reach over with his left hand and oppose his action.
>
> There is no place in the cerebral cortex where electrical stimulation will cause a patient to believe or to decide.[32]

Penfield therefore concluded that the mind or spirit was essentially independent of the brain. It used the brain, as a programmer might use a computer, but was not itself a function of the brain. Even though he had been a materialist for the forty or so years of his career, he ended his career as a dualist, as a result of his own research. His findings support the idea of the soul as a holistic cause, which influences the parts of the body, including the brain.

Contemporary neuroscientists would probably say that the ability to judge and decide are properties of the whole brain system (and therefore not located in any particular part of the brain). But if so, they function like a holistic cause: the whole affects the actions of the parts.

This is also evident in the mind's influence in healing, evident in the placebo effect and elsewhere. It is now widely accepted in the medical community that mental attitude affects health. If people think they are going to recover, they are more likely to recover than those who are pessimistic. Recently an article in the *Wall Street Journal* claimed that even thoughts can cause genes to turn "on" and "off." This article cited an anonymous writer in the journal *Nature* who noted the following. "I have had to spend periods of several weeks on a remote island in comparative isolation." He noted that the day before he was due to go on leave, his beard grew noticeably. "I have come to the conclusion that the stimulus for this growth is related to the resumption of sexual activity." What is remarkable is that it was his thoughts that triggered the release of testosterone, which stimulates beard growth. Developmental psychologist David Moore of Pitzer College in Claremont, California, is quoted as saying, "Just as stress in the med students I wrote about last week altered the expression of genes in their immune systems, so libidinous thoughts seem to affect gene expression."[33] Thoughts, then, can affect genetic expression, by influencing hormones that bind to the DNA, turning genes "on" or "off." This is dramatic evidence for the influence of mental activity on the biological matrix of the person, and indirect evidence for the influence of the whole on the parts.

If holistic causes do indeed exist at the level of cells, it would be reasonable that they might also exist at the level of complex organisms. The trajectory of contemporary molecular biology is to try to explain the whole by reduction to the action of the parts. But there may be room also for holistic causality. If the human soul is a holistic cause, whose effect as a field of active information is to order and direct the whole person, this would be a complementary cause to the part-whole causality that is investigated by the sciences. In the human person, then, there would be dual causality. The parts affect the whole, but the whole also affects the parts. The hypothesis of a holistic cause need not conflict with any of the work being done in contemporary neuroscience, which is principally focused on the influence of the parts on the whole.

How might the soul affect the matter of the body? The link between them is information. The neuronal networks of the brain are merely informational patterns embodied in the interconnection of the brain's neurons. It may be that the soul, as an active field of information, affects these neuronal networks at the quantum level or through some other mechanism. Such an effect could change the interconnections between neurons, and so change the pattern of the neuron networks themselves.

■ The Soul as Transcendent and Immortal

If the soul is an embodied active information system, how can it survive outside the body and the brain? My hypothesis is that the human soul is not naturally transcendent or immortal. It becomes transcendent and immortal through a divinely initiated gift of a personal relationship with God. (This hypothesis is similar to the theory of Catholic theologian Karl Rahner that God elevates the end or finality of human beings from a natural to a supernatural end, namely union with God, by what Rahner calls the "supernatural existential.")[34] By saying that the soul is not naturally immortal, I mean that its transcendence has not emerged naturally in the course of evolution, and further, that its transcendence and immortality go beyond nature as measured and tested by natural scientists. (Also, it is not immortal by its nature, as Plato seems to have thought.) According to the Christian Scriptures, only God and Christ possess immortality—"It is he [Christ] alone who has immortality and dwells in unapproachable light . . ." (1 Tim. 6:16). Simon Tugwell has argued that the idea that the soul is naturally immortal is not a Christian belief: God alone is the source of immortality.[35] God, at some undetermined time, perhaps sometime at conception, perhaps after conception, enters into a relationship with the newly formed person. This personal relationship elevates the capacities and horizon of awareness of the developing embodied soul or person, just as grace elevates human nature. Its relationship with God allows it to become aware of a transcendent Other, which Christians (and others) call God. This has scriptural warrant: Jeremiah writes "Before I formed you in the womb, I knew you, and before you were born, I consecrated you." (Jer. 1:5) As a result of its relationship to God, the person's horizon is elevated beyond that of the purely natural, material and social, beyond that of the animals, to the transcendent. This appears in human evolution with the fact that early humans begin to bury their dead with food and grave goods, indicating a belief in afterlife. There is no comparable evidence in the lives of animals, even the higher animals.[36] Another capacity of the soul that flows from its relation to God is freedom. For material systems by themselves are not free; only God is truly free, and God confers that freedom on the emerging person. And finally, God does not abandon his relationship with the embodied soul—for, as St. Paul writes, "the gifts and the calling of God are irrevocable" (Rom. 11:29). Therefore the soul can survive bodily death with at least some of its faculties intact. In this state it awaits the resurrection of the body. This position bears some resemblance to that of John Polkinghorne. He holds that the molecular pattern of any person survives in the mind of God, and is recreated by God in the resurrection, in an appropriate environment. "The soul is the form of the body. . . . The pattern that is me is dissolved at death, but it is a coherent hope that it will be remembered by God and reconstituted by him in his great act of resurrection."[37] Polkinghorne would deny that the soul exists as a separate entity after the death of the body, and his notion of the soul as a static pattern differs somewhat from my notion of the soul as a dynamic organizing principle. Yet his idea that the pattern of the deceased person survives in the mind of God, and my idea that God through his relationship confers immortality on the soul and holds it in being, may be similar.

Now, it might seem that my hypothesis is at odds with the traditional Christian position, which is that God creates each human soul from nothing. This was Aquinas's position and remains the official position of the Catholic Church.[38] My position, by contrast, seems to be a transformation rather than a creation *de novo*.

The problem with the traditional position is this: it seems to imply that God creates a thing, a new substance, and injects (or infuses) it into the new being—the zygote perhaps. But this sounds just like Cartesian dualism—a soul in a body, like a ghost in the machine. And we have seen that this position is today indefensible. But if the soul is the form, the organizing principle of the body, does it make sense to think of God as injecting a new formal principle? I think it makes more sense to think of God entering into a relationship with the new person, perhaps even at the moment of conception (though we cannot know this) and elevating the capacities of this new being. That is exactly how God works in other situations—grace, which builds on and elevates, but does not replace nature, and miracles, which work through nature and elevate it to a transcendent state, but do not supplant it (see next chapter). The only case where God seems to create something from absolutely nothing (as far as we can tell) is in the initial act of creation itself, the emergence of the first matter/energy in the universe.

Again, it is a misconception to imagine that God injects a new soul into the person without that soul being already related to God. For, as Norris Clarke has written, "To Be is to be substance in relation."[39] So the new soul, from the first moments of its being, would be related to God, as its true and final end. The new soul is not a "thing" exactly. It is perhaps better described as a "subsistent relation," the term Aquinas uses to describe the persons of the Trinity.

This could still be called a "creation from nothing," loosely interpreted. For the gift of the embodied soul's freedom, relation and awareness of God, and immortality is from nothing, and this is precisely what makes the person human, theologically speaking. We can imagine a proto-human to whom the gift of relationship with God was not given. Such a being might have some language ability, some intellectual capacity, but would not have any direct personal awareness of transcendence or God or inner freedom or immortality. His horizon and end would be nature, not transcendence. But in Christian theology, we would not call such a being "human," for humans are made in the image and likeness of God and therefore in relationship to God. Indeed, it is just such

176

proto-humans that might have preceded the emergence of fully human persons in evolution.

A similar process would occur in every human gestation. At some point in the gestation process, God establishes a personal relationship with the emerging human so that that person develops within a transcendent, and not merely natural, horizon. I am not convinced that this necessarily happens at conception, nor am I certain that it would have to happen in all cases where an egg is fertilized by a sperm.

This theory of the soul does not mean that the soul is fully formed at birth. The embodied soul, or whole person, grows and develops in self-awareness, skills, social relationships, knowledge, and wisdom, as has been described by child psychologists like Piaget and others. Genetics, environment, choices, and the grace of God are all factors in the development of every human being. As an embodied soul, the person is a psychophysical unity and makes choices that will determine its ultimate destiny: eternity with or absent from, God.

Finally, I am *not* arguing for immortality instead of resurrection, but for immortality *and* resurrection. Indeed, I think the resurrection of humanity at the end of time makes no sense if there is no continuity between the person who dies and the person who is resurrected. That continuity is the soul.

■ Is This Dualism?

Is this dualism? On the one hand it would seem so, for the soul, in my view, can survive the death of the body. I would agree with Aquinas that as separated from the body the soul is incomplete. Its nature is to inform a body, and it will not therefore be complete until the resurrection. But, if we can trust the evidence from near death experiences, the core of the personality, memory, and some forms of perception survive brain death.

On the other hand, I would contend that my position is not dualism. Explaining this will involve a short disquisition into the relation of God to the world. Modern science and most modern philosophy starts from matter as that which is most certain and explainable. Some philosophies then try to work their way from matter to God. But inevitably in this process, a dualism is introduced, usually between God, who is pure spirit, and the world, which is material. (Sometimes this dualism is between matter and the mind—as in Cartesian dualism.) One result of such a dualism is that it becomes very difficult to explain how God can act in the world, since, by definition, the world and God are radically different. What is lacking in contemporary approaches is any reality or concept that can bridge the gap between God and the world.

The problem is the starting point: matter, which is assumed to be certain and explainable. In fact, we do not fully understand what matter is. Modern science can analyze matter down to the level of quarks and gluons. But there still is no way to unify quantum theory and the theory of relativity. The best candidate for such a unification is so-called string theory. This theory proposes a level of matter even more fundamental than quarks and gluons: incredibly tiny one dimensional strings which vibrate in ten dimensions, only three of which are measurable by our instruments. So far string theory has proven to be too remote for experimental verification. So the roots of matter may well reach into areas of reality that go beyond that which can be detected with our instruments.

I propose that instead of starting with matter we start with God, who is, to use the language of Thomistic philosophy, absolute Being. God creates matter, but matter cannot be completely different and separate from God, else God could not relate to it at all. It is other than God but not wholly dissimilar. There is a difference, a great distance between God and matter, but also an underlying unity in being. If there were not, God could not relate to matter at all nor act within the material world. There would be an impassible gulf. So matter might be thought of as an expression of God that is finite, bounded, and governed by its own laws, but still, in some way, connected to God. It is a mistake to imagine that matter and God are two wholly different realities. If we think this, there is no way God can act in the world, and no way we can get to God from the world of matter. (For example, we have seen above that if mind is a result of a complex material matrix, the brain, then we cannot say how God could have a mind, since God has no physical brain.)

What then is the best term to designate the reality that is common between God and the world? In the philosophy of Aquinas, that common reality is termed "being," that is, the act of existing. However, the term "being" tends to be opaque to modern readers, so perhaps we could substitute for it the word "spirit," and think of God as spirit, and matter as finite, limited, crystallized spirit. Such an idea is central to the theology of the Catholic theologian Karl Rahner. "Matter," Rahner writes, ". . . is the outward expression and self-revealing of personal spirit, in the finite realm. Consequently, by its very origin it is akin to spirit, an integral factor for spirit and for the eternal Logos . . ."[40]

We have, then, a gradation or hierarchy of being (or spirit) from un-limited, absolute being (spirit) to limited being (spirit), namely matter. (Hierarchy here does not mean oppression or domination, but is used as in biology to indicate that higher levels include but go beyond lower levels.) Within this hierarchy, we can also locate the soul. Like matter it is finite, limited spirit, but it is not so constrained as material reality. It has its own independence from God but exists in a personal unity with God. It is finite but more conscious than pure matter—more free, more capable of self-direction and activity. Whereas matter is passive, the soul, as an active field of information, a holistic cause, actively informs and organizes its body. And it is linked with matter, for its nature is to inform or form matter. And so the human soul develops by its incarnation in the world of matter.

Another way to think of the relation between God and creation is through the analogy of information. God, who is omniscient, is the absolute fullness of information. God can conceive of all possible states of being, all possible universes. God is therefore infinite mind or infinite information. But information specifies form. So to say that God is the fullness of information is also to say that God can conceive of all possible beings and all possible states of being. Created, material beings on the other hand can be characterized as finite or limited information and hence as limited forms. A hydrogen atom is a specific, limited form, and can be specified by a relatively small amount of information. Cells are much more complex, and human beings even more so, but both could be specified by enough information (which would give, for example, the position of all their molecules and how these molecules interact with each other). So God is absolute information (containing all possible forms), and matter, limited information that has been instantiated or crystallized into a finite form or object. This is also true of the laws of physics and the physical constants, which are specified by information. But the soul, as holistic cause, is also a finite field of information, whose dynamic is to inform (or form) matter. In contrast to matter, which might be thought of as passive or static information, a soul is an active source of information that (to some extent) shapes or forms the matter of its body.

So, whether we use the language of being, or spirit, or the analogy of information, we can conceive of a hierarchy or unity between God and creatures who embrace God, the fullness of reality, souls or finite spirit, and matter. We do not have strict dualism, therefore, but a difference within a unity.

Now many will object that this makes God merely one more object within the field of being, but this is not true. If God is infinite, God cannot be an object or "a" being. Rather, in Thomistic language, God is Being Itself, unlimited being, whereas all creatures are limited and finite being. God exists *per se*—God cannot not exist, whereas every creature may or may not exist; creatures exist contingently. So there is a great distance, even perhaps an infinite distance, between God and creatures.[41] But God is not "wholly other."[42] For if God were wholly other, there could be no communication between God and the world, in either direction, even by revelation. For God to reveal himself to embodied creatures, there has to be some underlying similarity or unity between God and creatures. Nor is it true to say that this makes Being greater than God, with God being only one instance of being. God is the infinite fullness of Being (or spirit or information) who holds all finite beings in existence.

This, therefore, is not dualism, nor is it monism. Perhaps it could be called dipolar monism, providing that we recognize that the two poles of the monism, God and matter, are in no way equal. I realize that it will seem to naturalists to be merely mythology. But at least it can account for the reality of God, spirit, soul, and matter, along with free choice, mental causation, and afterlife, all within one conceptual scheme with no unbridgeable dualism at any point.

■ Conclusion

Naturalism, as represented by Francis Crick, E. O. Wilson, and others, reduces the human person to the sum of his or her molecules, or perhaps the sum of his or her genes. Wilson, for example, writes: "In a Darwinist sense, the organism does not live for itself. Its primary function is not even to reproduce other organisms, it reproduces genes, and it serves as their temporary carrier. . . . [T]he organism is only the DNA's way of making more DNA."[43] Later in the same book he opines: "Altruism is conceived as the mechanism by which DNA multiplies itself. . . . [S]pirituality becomes just one more Darwinian enabling device."[44] In such a worldview it is hard to find any basis for human dignity, inalienable human rights, free will, mental causation, God, or transcendent moral values—indeed, transcendent anything.

Emergentism, in its various forms, tries to affirm these realities while also affirming the physicalist assumptions of contemporary science—that is, that within the natural world everything has a physical cause, including the mind. Many Christians espouse this view. But while emergentism is correct up to a point, it eliminates any basis for belief in an immortal soul, and therefore calls afterlife into question. Emergentists may take refuge in a belief in the eventual resurrection of the dead at the end of history. But emergentists typically want to affirm that there are no breaks in nature in the chain of physical causality (except perhaps at the level of quantum indeterminacy). That is, every event can be explained by physical causes. But this seems to eliminate miracles, and the bodily resurrection, if it is anything, is a miracle. Therefore I do not see how the bodily resurrection is coherent with an emergentist position. I also think it is not sufficient to provide a solid ground for our experience of human freedom, our awareness of the presence of God, or for the mind and consciousness of God.

For these reasons I have advanced, as an hypothesis, the notion of the soul as a holistic cause, the form of the body, a position deriving from that of Thomas Aquinas. In such a view, the parts of the person affect the whole. Thus there is no contradiction with the findings of biology and neuroscience. But the whole also has a (limited) effect on the parts—the soul is the organizing principle of the body. It thus complements the part-whole causality which is investigated by reductionistic science. I also think a plausible case can be made that the soul, because of its relation to God, can survive the death of the body. But this survival is a gift of God's grace; the soul is not naturally immortal, as in Platonism or Cartesianism. Finally, if one starts with the material world as that which is most certain and reliable, there is no good way to get to the world of spirit and God. There is a gulf, a dualism between these two realms. But if one starts with God, conceived of as Being, spirit, or information, then one can understand the material world and the soul as limited and finite being, spirit, or information—that is, as the creation of God. And so there is no irreconcilable dualism between God and the world.

The next chapter will explore one of the consequences of this position: the possibility of miracles.

8

Miracles
Signs of God's Order

■ The belief that everything can be explained by natural causes, and that natural causes are those that science can detect, almost by definition excludes any belief in miracles. Historically, the first target attacked by those tending to naturalism, the Deists, was the miraculous. Lord Herbert of Cherbury, for example, argued against supernatural intervention, especially miracles, but also against revelation. The Deists contended that God could be known by reason through nature, and that anything that contradicted reason, such as the Trinity, the incarnation, or miracles, was superstition. Thus Matthew Tindal, in his *Christianity as Old as the Creation: or, the Gospel, a republication of the Religion of Nature*, wrote, "Nothing can be requisite to discover true Christianity and to preserve it in its native purity free from all superstition, but after a strict scrutiny to admit nothing to belong to it except what our reason tells us is worthy of having God for its author."[1] The Deists, then, were rationalists; anything that struck them as unreasonable, or that did not fit their conception of nature, was rejected, typically without any examination of the evidence.

The Deist attitude toward miracles was carried forward into the eighteenth century by Enlightenment thinkers. The most famous of these was David Hume, who defined a miracle as "a violation of the laws of nature."[2] This definition seems calculated (as in fact it was) to turn people against the very possibility of miracles. For what many persons (especially scientists) find so beautiful about nature is precisely its lawful complexity. As Paul Davies said in his Templeton Prize address, "To me, the true miracle of nature is found in the ingenious and unswerving lawfulness of the cosmos, a lawfulness that permits complex order to emerge from chaos."[3] Nothing could be more repellent, both scientifically and theologically, than a God who capriciously intervenes to heal this person while ignoring that, or who arbitrarily interrupts natural processes to produce a "wonder" to astonish the multitudes. As Arthur Peacocke writes:

> [I]ntervention suggests an arbitrary and magic-making agent far removed from the concept of One who created and is creating the world science reveals. That world appears increasingly convincingly as closed to causal interventions from outside of the kind that classical philosophical theism postulated (for example, in the idea of a "miracle" as breaking the laws of nature).[4]

In this chapter, I will argue that the idea of miracle as a "violation" is misleading, and will advance the notion of miracle as an event, caused partly by God, which is consistent with, but yet transcends, natural processes. This understanding of miracle is more in keeping with both good science and good theology. Against naturalism, I will maintain that miracles do occur. They are signs that nature is not all there is, but that it exists within a more comprehensive, divine context. This context makes its presence known not only through nature's lawfulness, but also through miraculous events that reveal the existence of a higher order.

In traditional theism (Judaism, Christianity, and Islam as well as some branches of Hinduism), miracles have been understood theologically as signs: events in which divine action is unusually apparent, and which testify to divine authorship. The miracles worked by Yahweh through Moses in Egypt were done so that "The Egyptians shall know that I am the LORD" (Exod. 7:5). Jesus' miracles, especially in the Gospel of John, "revealed his glory; and his disciples believed in him" (John 2:11). In John 3, Nicodemus says to Jesus: "Rabbi, we know that you are a teacher who has come from God; for no one can do these signs that you do apart from the presence of God" (John 3:2). Muslims also accept some of the miracles of Jesus, but consider that the greatest miracle is the Quran itself, which could not have been produced by an illiterate person (as Muhammad was) and so must be the words of Allah.

What challenged this traditional perspective on miracles was the rise of modern science and its view of nature. Nature was no longer seen as a system open to divine action, as it was in traditional Western and Eastern religious thought. Rather, it became understood as a more or less closed system, in which God could only intervene from the outside, as it were, as a mechanic would intervene to alter the workings of a clock or watch (a favorite metaphor of the Deists). This led to Hume's definition of a miracle as "a violation of the laws of nature." Defined in this way, miracles came to be seen as immensely improbable—so improbable, in fact, that, as Hume argued, any reported miracle would always be less probable than the possibility that the laws of nature had been violated. Miracles became, therefore, effectively impossible by definition. Furthermore, implicit in the very idea of a miracle as a "violation" is the sense that miracles are irrational and outside of any lawlike order, a notion deeply repugnant to scientists.

For these reasons miracles have been largely ignored in modern science and in discussions of science and theology. And this has been due to an *a priori* attitude that pays scant attention to any contemporary evidence for miracles (which is in fact extensive). Miracles are dismissed as inherently impossible or unworthy of attention, both on scientific grounds and on theological grounds. Thus, an unsigned editorial in *Nature* in 1984 stated: "Miracles, which are inexplicable and irreproducible phenomena, do not occur—a definition by exclusion of the concept."[5] And the famous German theologian Rudolf Bultmann asserted that miracles were "mythology." In a passage that would have pleased the editors of *Nature*, he wrote: "Modern men take it for granted that the course of nature and history . . . is nowhere interrupted by the intervention of supernatural powers."[6] But neither the editors of *Nature* nor Bultmann felt it was necessary to actually investigate evidence for miracles. We know, they assert, that miracles cannot happen; therefore they can be dismissed *a priori* as impossible.

Yet, if the hallmark of empirical science is impartial openness to evidence, such ways of proceeding can hardly be called scientific. If scientific laws are held to be not only descriptive but also prescriptive, telling us what can and cannot occur in nature, then we would have to dismiss all evidence that cannot be explained by present scientific theory. Thus recent experimental evidence that has established that widely separated "entangled particles" interact nonlocally (or else communicate at faster than light speeds) would have to be rejected, since it is inexplicable according to current scientific theory.[7] A letter by Donald MacKay (Department of Communications and Neuroscience, University of Keele, U.K.), responding to the editorial in *Nature*, explains this well:

In scientific laws we describe, as best we can, the pattern of precedent we observe in the sequence of natural events. While our laws do not prescribe what must happen, they do prescribe what we ought to expect on the basis of *precedent*. If by a "miracle" we mean an unprecedented event

... then science says that miracles ought not be expected on the basis of precedent. What science does not (and cannot) say . . . is that the unprecedented does not (or cannot) occur. . . . We cannot dogmatically exclude the everpresent possibility that the truth about our world is stranger than we have imagined.[8]

I will argue here for four points. (1) The evidence for miracles, ancient and modern, is respectable and deserves attention; it is unscientific simply to ignore it. (2) Hume's definition, which has monopolized modern discussions of the miraculous, is misleading, both scientifically and theologically. (3) Miracles are better understood as signs of divine action that, like grace, do not violate nature but work through it, perfect it, and reveal its divine ground, for nature is not a closed system but an open system within a larger, divine context. Viewed in this context, miracles can be seen as rational and even lawlike events that express the divine ground within which nature exists. Just as the laws of nature behave in extreme ways in unusual contexts (e.g., superconductivity or black holes), so it may be that the ordinary laws of nature, within a context of faith, behave in unexpected ways. (By a context of faith I mean one in which persons of faith and holiness relate to God in prayer.) (4) A consideration of miracles has something to contribute to the contemporary discussion (prominent in science and theology dialogues) about the nature of divine action in the world. I will not argue here for theism, or for the existence of divine action, but only that, if God exists and does act in nature, by what theologians call special providence or special action,[9] miracles can be understood as instances of this in a way that makes sense theologically and scientifically.

■ Evidence for Miracles

Let us begin with some evidence for miracles. There is a great deal of scriptural testimony for miracles: almost one-third of Mark's Gospel, for example, is made up of miracle stories. But these stories are so distant in time, so scantily described, and in some cases so laden with symbolism (e.g., the water into wine, John 2) that it is hard to tell exactly what happened. A basic axiom of science and historiography is that we assume that the same natural laws and kinds of events that are operative today were operative in the past. Thus, if we can find well-attested miracle stories in contemporary times, they would lend support to the ancient accounts. Let us then consider some modern accounts.

A famous example comes from Dr. Alexis Carrell. In 1902 Carrell, a young surgeon from Lyons, traveled by train to Lourdes to investigate reports of extraordinary healings. At the time, he did not believe in the possibility of miracles, but a few of his patients had returned healed from Lourdes. Carrell, as a good scientist, wanted to investigate. During the journey, he was asked to care for a young girl, Marie Bailly, who was dying of advanced tuberculosis. Carrell examined her and confirmed the diagnosis; her abdomen was badly distended and hard, her heart racing; she was near death. He described her case to a friend on the train:

> This unfortunate girl is in the last stages of tubercular peritonitis. I know her history. Her whole family died of tuberculosis. She has had tubercular sores, lesions of the lungs, and now for the last few months a peritonitis diagnosed both by a general practitioner and by the well-known Bordeaux surgeon, Bromilloux. Her condition is very grave. . . . She might die at any moment right under my nose. If such a case as hers were cured, it would indeed be a miracle. I would never doubt again; I would become a monk![10]

Upon reaching Lourdes, Ms. Bailly was carried on a stretcher to the Grotto, but not immersed in the pool, since the attendants feared it would kill her. Carrell meanwhile kept a detailed record of his observations:

> Condition very serious . . . Abdomen very sore and distended. Pulse irregular, low, almost imperceptible at 160; respiration jerky at 90; face contracted very pale and slightly cyanosed. Nose, ears, extremities cold. Dr Geoffray of Rive-de-Gare arrives, he looks at the girl, feels, strikes, and ausculates the heart and lungs and pronounces her in the last agony. As there is nothing to be lost and the patient wishes to go to the Grotto again, she is taken there on a stretcher.[11]

And yet at the Grotto, she was healed. Carrell and several other physicians examined her carefully at the Lourdes Medical Bureau (where their signed attestations are on record). The physicians' report reads (in translation):

> May 28, 7 P.M. Our stupefaction was profound on seeing the girl who was so ill this morning sitting on her bed, chatting with the nurses and answering our questions smilingly, also at seeing the enormous swelling of the abdomen had disappeared. The tumors which had encumbered it had melted away visibly; respiration and heart had resumed their normal play. This is a sudden, wonderful cure, a veritable resurrection.[12]

Dr. Geoffray added in his own hand: "This medical authentication which I am signing is the simple truth; so grave an affection [sic] has never been cured in a few hours like the case on record here."[13]

Some months after her cure, in perfect health, Marie Bailly joined the Sisters of St. Vincent de Paul. Carrell, though dumbfounded by the cure, did not become a monk. But because of his interest in miracles, he was told that the Lyons Medical School would never open its doors to him. He left for America to join the Rockefeller Institute and in 1912 was awarded the Nobel Prize in physiology and medicine. His recollection of Marie Bailly's cure was published as *The Voyage to Lourdes*.

A second case, also from the annals of Lourdes, is that of Joachime Dehant, a 29-year-old Belgian woman, who in 1878 made a pilgrimage to Lourdes by train. She sought healing for a massive gangrenous ulcer (12 inches by 6 inches) on her right leg, an ulcer that had penetrated to the bone, destroying muscles and tendons, and caused the foot to invert. The wound had been attended by local physicians for twelve years, but it had not healed; it had been years since Joachime had walked or worn a shoe on her right foot. Her trip to Lourdes was agonizing and humiliating: the stench from her gangrenous wound so sickened her companions on the train that they vomited. Yet on her second bathing in the pool at Lourdes, the wound was healed; muscles, tendons, and skin were restored, leaving a well-formed scar. This cure was attested by physicians at Lourdes, by her traveling companions, by the townspeople of Gesves, Belgium (Joachime's home), who had known her for years, by the physicians who had treated her for twelve years at Gesves, and by her family. It was subsequently investigated by professors at the Catholic University at Louvain, who gathered testimony from all concerned, including physicians. Joachime enjoyed perfect health for more than thirty years after the cure.[14]

There are hundreds of stories like this, scattered through Christian literature as well as the writings of other religions, such as Hinduism.[15] Of course, many such stories can be explained by coincidence, wishful thinking, autosuggestion, fraud, superstition, and so on. Everyone knows of accounts where a crippled person flings away his crutches at the behest of the faith healer, only to collapse in a worse condition later. But amid all the chaff there are some grains of wheat: some accounts are so well attested, by responsible persons (sometimes scientists) who have everything to lose and nothing to gain by testifying (like Carrell), that it is difficult to lightly dismiss them. And many of the cures concern deep-seated organic diseases that had resisted treatment for years, and yet the cures were sudden and permanent. At least within Roman Catholicism, no event will be claimed to be miraculous if it can possibly be explained by natural causes. This means that part of the process for discerning a miracle is a thorough scientific investigation. Only after such an investigation has concluded that there is no scientific explanation for the event can it be declared a miracle, and then, only if it has occurred in a context of faith and prayer. My claim, then, is simple: such evidence deserves our attention and should not be lightly dismissed because we are convinced on *a priori* grounds that miracles cannot happen. How they might be explained is another and subsequent question; the first obligation of any dispassionate investigator is to be open to the evidence. Then we can consider causes and explanations.

■ Are Miracles Violations of Nature?

What about Hume's contention that miracles are a "violation of nature"? There are some interesting points to notice about both the accounts given above. First, it appears that neither of the healings was "instantaneous" (a term used too readily in describing miracles). Each of them appears to have taken some time: perhaps a half hour in the case of Dehant (she was in the pool twenty-seven minutes, and experienced great pain, but the wound was not sore on leaving); perhaps two or three hours in the case of Bailly. Second, in Dehant's case, the healing terminated in a pronounced scar. Now both of these facts, but particularly the second, suggest that the healing was not instantaneous, but was rather a greatly accelerated natural process. And this is borne out by other observations. In an article, Alexis Carrel writes, "I believe in miraculous cures, and I shall never forget the impact I felt watching with my own eyes how an enormous cancerous growth on the hand of a worker dissolved and changed into a light scar. I cannot understand, but I can even less doubt what I saw with my own eyes."[16] Carrell concluded that: "The miracle is chiefly characterized by an extreme acceleration in the process of organic repair."[17] Many other observers have also stated that cures take place as natural processes, but at much greater speeds.[18] Such accounts, then, indicate that if miracles are the result of divine activity, such activity does not "violate" nature or work around it, but rather works through it to heal. In this respect, its action in miracles is similar to its action in what theologians call grace, which does not work against nature, but builds on nature, healing and perfecting it. This explanation also follows Christian tradition. R. M. Grant writes: "In the first five centuries there is a tendency to explain God's working in miracles as beyond or above nature but coordinated with nature rather than contrary to it."[19] Medieval authors also, such as Aquinas, usually described miracles as *praeter naturam* (beyond nature) rather than as *contra naturam* (contrary to nature).

However, there are certainly biblical miracles that seem to go far beyond nature. Such are, for example, the changing of the water into wine (John 2), Jesus' walking on water (Mark 6:47–52), and the resurrection of Jesus (Matt. 28:1–10 and parallels). Some of these are weakly attested (the conversion of the water into wine is found in only one Gospel), but others, like the resurrection, are found in all four Gospels. How can these be explained?

They are, first of all, extreme examples, scarcely encountered outside of Scripture. I know of no resurrection stories or water into wine stories in modern literature on miracles. However, I do accept the resurrection of Jesus on the basis of Paul's testimony in 1 Corinthians 15, the dramatically changed behavior of the apostles, the fact that opponents of Christianity were never able to produce Jesus' body or even a burial site, the fact that women were the first witnesses of the risen Jesus, and for other reasons (for a defense of the historicity of the resurrection by a scientist, see Polkinghorne, *The Faith of a Physicist*).[20] But an explanation will involve a theological digression.

If one believes in the God of Judaism, Christianity, and Islam, or even in the Brahman of Hinduism or the Sunyata of Buddhism, then the natural world, as described by science, is not all there is to reality. Rather, the universe exists in a wider context, which theists call a divine context. God transcends nature and holds nature in being, so that we might say (to use a spatial metaphor) that nature exists in God. Thus nature is not a system that is closed to the possibility of divine influence, but is open to it. This is widely accepted in theology-science discussions, but I will briefly give a few reasons here, and refer the reader to other literature. First, many have argued that the indeterminacy evinced in quantum mechanics is an ontological indeterminacy in nature itself, and that therefore God can act to specify quantum indeterminacies. John Polkinghorne argues that chaotic systems are so exquisitely sensitive that the tiniest influence (perhaps even at the quantum level) can cause significant changes in macro level systems.[21] Brian Grene, a physicist, argues that string theory is the only theory that can unify both gravitational theory and quantum theory, and that string theory postulates that matter is formed of strings which vibrate in nine spatial dimensions. This means that even the roots of matter itself reach past what can be observed in physics, and that the nature of matter itself is not yet fully understood.[22]

Now, if nature is understood as an open system within a context of divine activity, it may be the case that in some extreme circumstances, such as the presence of great faith, the laws of nature, while not changed, behave differently than in ordinary contexts. (Superconductivity and black holes are examples of the laws of nature behaving unusually in unusual contexts.) Ordinarily, divine activity works through and in coordination with the laws of nature, usually invisibly, but sometimes strikingly, as in the cases of extraordinary healings. But in very rare or

190

unique events, such as the resurrection of Jesus, nature becomes, as it were, transparent to its divine ground, and behaves in extraordinary ways. But even in the resurrection, if we can trust the biblical accounts, the divine activity respects the laws of nature: the body of Jesus is transformed and appears to be in a different context, freed from the limitations of our space and time, but is still recognizably the same body as that of the crucified Jesus.

Some analogies may help to clarify this point. It is commonplace in Catholic and Orthodox theology to speak of human nature as being elevated by grace. Consider love, for instance. Human beings have some natural capacity to love, but it is limited; loving one's enemies seems to be beyond our natural capacities. But the love of God, poured into our hearts by the Holy Spirit (Rom. 5:5), can elevate this love so that it is no longer self-centered (we love those who love us) but universal, as is God's love. Something similar may be possible in natural processes. Just as the human psyche or will can transcend its normal state through the action of grace, so perhaps can the normal healing processes of the body transcend their usual capacities through the action of miraculous grace. But both the normal state and the transcendent state are states of nature. Let us consider another analogy. Ordinary matter can exist in several states—solid, liquid, gas—but also, at much higher temperatures, as an atomic plasma, in which the electrons are stripped from the nuclei. Somewhat different physical laws apply in each of these states: the laws of chemistry, for example, would not apply to the atomic plasma state, because the atoms do not have orbiting electrons and cannot form chemical bonds. Physicist Paul Davies notes that ". . . it was discovered long ago that the law of conservation of matter is violated when new particles and anti-particles are created in high energy processes; . . . It has been suggested that if one could drive the energy up without limit, one by one all of our cherished laws would fail. . . ."[23] So also, what we call "nature" may have the capacity to exist in more than one state, a ground state, normally investigated by science, but also perhaps "graced" states in which it is elevated by the presence of miraculous grace. Such may be the case in miraculous healings or in the resurrection. But even in such cases we are dealing with potentialities of nature. The resurrected body of Jesus (if it is/was really the same body that died, but in a transformed state, as traditional Christianity affirmed) is also within the capacities of nature, but a nature transformed by grace or divine activity.

If this idea is correct, it means that miracles are never *only* the activity of God, as if God in a miracle acts alone, nakedly, in place of natural causes. Theologically, a miracle, even an exalted miracle such as the resurrection, is always God working through or in cooperation with nature, and not against it. Augustine puts this idea well:

But God, the Author and Creator of all natures, does nothing contrary to nature; for whatever is done by Him who appoints all natural order and measure and proportion must be natural in every case. . . . There is, however, no impropriety in saying that God does a thing contrary to nature when it is contrary to what we know of nature. For we give the name of nature to the usual common course of nature; and whatever God does contrary to this we call a prodigy, or a miracle. But against the supreme law of nature, which is beyond the knowledge of both the ungodly and of weak believers, God never acts, anymore than He acts against Himself.[24]

Now this suggests some obvious objections. If we really know so little about nature, how can we be sure that miracles are caused by divine activity, and not by some yet-unknown law of nature? I don't think we can be certain, at least on the grounds of reason. Belief in miracles involves faith as well as reason. A scientist, speaking as a scientist, can say: "There does not seem to be any known natural explanation for this event." But as a scientist she cannot conclude that it is therefore a miracle. That is a subsequent step, involving faith. It is possible, of course, that what has hitherto been regarded as miraculous will one day be explained when we have a more complete understanding of nature. But it is worth pointing out that the kinds of events that have been traditionally labeled miraculous—sudden and permanent healings of organic diseases, levitations, the resurrection of Jesus, and so on—are no more explainable now, after centuries of scientific advance, than they were in ancient times.

There are other reasons for doubting the "unknown law of nature" hypothesis. First, putatively miraculous events seem to occur within contexts of faith and individual or communal prayer. That is why they are more common in the lives of holy men and women. (This is also true of the resurrection, whose context is the whole life of Jesus, a life surrendered to faith and prayer.) But if they were due to an unknown law of nature, that law ought to be visible in secular contexts as well. The simplest explanation seems to be that these events involve divine action.

A variant of the "unknown law of nature" hypothesis is that healings, in particular, are due to some unknown capacity of the mind. An appeal to an unknown, however, does not constitute an explanation. Yet I would argue that if God does indeed work through nature and not around it, it is likely that the mind is involved in healings—how could it not be? But again, since dramatic healings of long-entrenched organic diseases seem to occur mainly in faith contexts, it makes sense to think that the best explanation is that somehow, by a mechanism we do not yet understand, divine activity, working through the mind, empowers the healing capacities of the human being in a way that transcends its usual capabilities. This, however, would only explain some miracles, not all; it would not explain the resurrection.

I have said that miracles occur within a context of faith and prayer. From a theological perspective, there are two possible ways of understanding how God might act in a miracle. The first is the traditional way: God responds to prayer, faith, and holiness. If a person or group of persons of holiness and faith pray to God for a healing, God may respond. A second way is this. Perhaps God's activity, or "energy," to use a modern analogy, is always and everywhere available, like an extended field or supporting context. Wolfhart Pannenberg has written that the Spirit of God might be viewed (analogically) as a dynamic field.[25] But if so, it is a field that can only be accessed by those who open themselves to God in faith, holiness, and prayer. An analogy here might be tuning into a radio broadcast. The radio waves are always present, but we are unaware of them unless we have a receiver tuned in to the proper frequencies. The first model envisions God's action in terms of personal response. But the second represents it as a field or context phenomenon; the field is always present, but only some access it. I think that *both* of these models are necessary to understand miracles, just as both particle and wave models are necessary to understand subatomic particles. The models are complementary, and either without the other is incomplete. The first model explains the fact that many miracles do seem to be responses to prayer. But the first model by itself is open to the objection: Why doesn't God heal everyone who prays? It may be that the reason is that to access the divine energy, it is necessary to be surrendered to God in faith and prayer, and that few people are. It is not that God plays favorites and rewards those who grovel the most. It is that those who are not deeply surrendered to God cannot access his power because they are not tuned in. For God to act in our lives, we have to be receptive, and if we are not, God cannot act as fully as God might otherwise. But if there really is divine activity in miracles, can we explain how it influences physical processes? What, in other words, is the "causal joint" between divine and physical activity?

I do not think at present we understand the mechanism. It might be that God acts at the quantum level. Quantum states, which are indeterminate, are determined by divine activity so as to influence physical processes. Robert Russell has proposed this model of divine activity as a way of explaining theistic evolution and special providence.[26] This might account for an accelerated healing. It is hard to see, however, how it could account for more dramatic miracles like the resurrection. My own belief is this. Theologically, and even logically, God cannot be completely separate from the created order. If God were "wholly other,"

God could not influence the world, nor could the world influence God. But this is not the Christian idea of God. Rather, it is the Deist idea, a consequence of viewing the universe as a self-enclosed mechanical system that leaves God on the outside. In traditional Christian theology, the theology of the Greek fathers and of Augustine and Aquinas, God is other than creation; indeed, God transcends creation infinitely, but there is also an analogous unity in being (chapter 3). As Aquinas expresses it, God's essence is to exist; God is the act of existence from which all other existent things derive their existence. There is therefore a unity, as well as a discontinuity, between God and creation. If so, (finite) spirit (such as the soul) could influence matter directly, and God, in turn, could influence the soul (this is how Aquinas explains the resurrection). I do not think, though, that God ever acts as one force alongside other physical forces, but rather acts in creation mediately, so as to empower nature to transcend itself.

But this leads to a further point, namely, that we don't yet completely understand what matter/energy is. Grene's book on string theory indicates as much. Its roots may extend past the boundaries that can be measured by physics, perhaps into higher dimensions.[27] Therefore at present we cannot say exactly how divine activity might influence matter; perhaps when we have a better idea of what matter is, and what its boundaries are, we will be in a better position to explain, however sketchily, how the divine activity in miracles actually works. Or perhaps such knowledge will be forever beyond our ken. But the fact that we cannot presently explain *how* a process works does not mean *that* it doesn't work. Newton and his successors could not explain how gravity worked, but that did not stop them from believing that gravity existed. No one knows right now how to explain the nonlocal interactions of so-called "entangled particles." The phenomenon, well established by experiment, seems to be completely beyond what we presently know of the laws of nature. If we followed Hume's dictum, we should say that the regular workings of nature are so well understood that these reports of nonlocal interactions must be spurious, and we would reject the evidence. But we don't do that; we accept the evidence that something inexplicable happens in nature, and we hope to be able to explain it sometime in the future, perhaps with a modified understanding of natural laws.

Miracles and Naturalism

What, then, is a miracle to a scientist and to a theologian? Miracles might, of course, simply be illusions: events that are really fabrications, or are coincidences, or are due to some mysterious power of the mind, or to an unknown law of nature, but are not due to any divine activity. In other words there are no miracles, theologically speaking; there are only unusual events. This, however, is a hypothesis that remains to be proven. But if part of the cause of a miracle is divine activity, then, to a scientist, a miracle will appear simply as an inexplicable event, a mystery, which, like the healings observed by Alexis Carrell, seems to go beyond what can be explained by natural causality. Natural causes will still be involved in the event—the healing of Joachime Dehant terminated in a scar, indicating that the healing followed natural pathways. But it cannot be explained completely by natural causes. To a theologian who believes in miracles (and there are theologians who do not), there is a two-step process in discerning a miracle. First, can it be fully explained by natural causes? Second, did it take place in a context of faith and prayer, as an apparent response to prayer? If the answers are "No" to the first question and "Yes" to the second, it may be a miracle.

This means that in the discernment of a miracle, the work of both the scientist and the theologian is indispensable; we need both science and theology to understand the miraculous. No responsible theologian is going to object to scientific investigation of a miracle (or miracles generally). The claim of miracle should never be used to block attempts at scientific explanation, to claim that this class of events is off limits to scientific analysis.

Finally, if the traditional claim is true that in miracles divine activity is unusually apparent, or, to put it another way, that in miracles nature is unusually transparent to its divine ground, then miracles can be seen as similar to laboratory experiments, which isolate one causal factor so as to study it apart from other factors. If this is so, miracles ought to be of interest to all those who are trying to understand how God acts in the world.

What are the implications of this notion of miracles for naturalism? Naturalists hold that everything can be explained by natural causes alone; there are no supernatural forces or causes. A maximal form of naturalism holds that only matter and the laws of nature exist—there is no God. A minimal form of naturalism holds that God exists, but that within the natural world, the chain of natural causality is never broken. David Ray Griffin advances this position in *Religion and Scientific Naturalism*,[28] and holds that it is a necessary assumption of natural science (I will consider this in the next chapter). But the possibility of true miracles, that is, of supernatural causality acting within the context of nature, as I have argued, is inconsistent with both these forms of naturalism. My contention is that nature is a system open to the action of divine grace and the miraculous, as I have described it above. The degree to which this occurs depends largely on our openness to God in faith. It is not the case that God arbitrarily decides to intervene here and not there, now and not then. The pattern in the Synoptic Gospels is that Jesus' miracles are a response to faith. Where there is little faith, Jesus cannot work miracles. The most striking example of this is when Jesus returns to Nazareth, his hometown, and can work "no deed of power there" because of their unbelief (Mark 6:1–6). (It is also true, however, that miracles strengthen faith. This is the interpretation of miracles prominent in John's Gospel.)

This means that nature exists with a transcendent order and is capable of being transfigured or elevated by that order, both in the act of grace and in the act of the miraculous. One characteristic of this order is love and personal fidelity. Those who are faithful to God, and who put their trust in him, will not be put to shame, but will be saved by God. And this higher order includes and supports, but also transcends, the order of nature. A familiar example of this principle is personal choice and volition. If I am hungry, I can choose to eat. That choice will be expressed through my body. But I can also choose to resist my bodily need for food, and fast. If the soul is the form of the body, it can go along with bodily urges, but it can also transcend them.

The prime example of such transcendence is the resurrection of Jesus. The Gospels describe it as a bodily resurrection. That is, his body disappears from the tomb, is transfigured so that it is no longer bound by our space and time, and is eventually exalted ("taken up") into heaven. Yet it is the same body as that of Jesus on earth, as is shown by the fact that it possesses the same wounds, even in its risen state. And Jesus' resurrection is a foretaste of the resurrection that awaits all the just. Furthermore, Paul insists that creation itself will be redeemed and transformed (Romans 8). So, the Christian theological evidence, at least, is that nature itself is capable of existing in an exalted state, through grace. And miracles are signs of this possibility, because, if my argument is correct, this is also what happens in miracles.

Such an interpretation made sense within the worldview of a hierarchy of being. But it makes no sense within a naturalistic worldview that sees reality as existing on only one level, that of the physical. Nor does it make much sense in a worldview that holds that reality consists of only two levels: the physical, and God. For it to make sense, there has to be at least the possibility of exalted states of being, and even exalted states of materiality, intermediate between our limited space-time universe and God. This is what the old hierarchy of being allowed for, for example in the possibility of angels, or in what St. Paul referred to as a "spiritual body" (1 Cor. 15:44).

Recent writing about the existence of higher dimensions introduces interesting possibilities for recovering a hierarchy of being that is credible within a contemporary worldview. Suppose one asks, "Where is the resurrected body of Jesus now?" Clearly, it is not within our space-time universe. One frequently suggested possibility is that it exists in a higher dimension. If so, it could appear and disappear within our space-time at will, but would not be limited to our space-time, as we are. (In the same way, I could project an image of myself onto a plane so that it would be apparent to a figure within that plane. But I am not limited to the plane, as is the figure within it.) So far, the existence of higher dimensions is controversial within physics, but it is becoming more accepted. String theory demands the existence of higher dimensions. And a recent article in *Scientific American* argues that scientists will soon be able to demonstrate the existence of higher dimensions.[29]

■ Conclusion

Miracles, then, are not capricious interventions by God. This is a caricature, which results from the conception of nature as a closed, mechanical system. But nature is an open system, existing within a hierarchy of being, and therefore open to transcendent activity of a higher order, God. This activity is consistent with the laws of nature, as in the acceleration of natural healing processes, which is characteristic of healing miracles. It is true, as Paul Davies asserts, that nature's lawfulness manifests the lawfulness and intelligence of its Creator. But if we stop there, we end with the God of Deism: an absentee God who creates the initial laws and constants and energy of the universe, but does not thereafter act within it, or can only act very indirectly. This is not the God of Christianity, a God who loves his creatures and out of that love acts within nature to help them. Much of the Gospel narratives concern miracles worked by Jesus, and to reject them is to reject a good deal of the gospel.

The miracle par excellence is the bodily resurrection of Jesus. It is this that vindicated Jesus' claims to be sent by God. But if miracles do not occur, neither did the resurrection. It is inconsistent to claim that miracles do not occur, or that miracles do not transcend natural laws, but to claim that the resurrection did.

Miracles are signs (not proofs) that nature exists within a higher order, an order of love and the fidelity of God. Just as divine grace heals and elevates the human intellect and will, so miracles elevate the activities of physical nature. They thus also reveal the potentiality of nature to exist in transcendent states, not suspected by naturalism. In a miracle, nature becomes transparent to its divine ground, like a window opening onto a higher state of being. Another way of putting this is that miracles are like sacraments: they are visible events in which the divine presence shines forth. If we believe not only in the resurrection of Jesus, but in the universal resurrection of the dead and the consequent transfiguration of nature itself, promised in Romans 8 and in Revelation 21:1 ("I saw a new heaven and a new earth"), then we can look forward to a heaven in which all of nature becomes a miraculous sacrament showing forth the presence of its Creator. Matter itself will be transfigured so that it does not occlude the divine presence, but expresses it. As John Polkinghorne writes: "The ultimate destiny of the whole universe is sacramental."[30]

Christianity and Science
Conflict or Complementarity?

■ Natural science is a powerful method of exploring the physical aspect of reality. But since it depends on experiments to test its hypotheses, it is also limited, by that very fact, to the realm of physical reality. It cannot tell us much about purpose and meaning, which values are good or bad, or whether there is a reality that transcends nature or spiritual entities such as angels or souls. Traditionally, the discipline that has claimed to provide knowledge about such things has been theology. Therefore if we desire knowledge of both the physical and spiritual aspects of reality, we will need both science and theology. In this sense, science and theology are complementary—each completes the other. This has been a theme stressed by John Polkinghorne in his many writings. But it was also a prominent theme in the writings of early modern scientists, many of whom were Christians. Galileo, for example, spoke of the book of nature and the book of Scripture as two aspects of God's revelation. Not everyone, however, agrees that science and theology are complementary; furthermore, "complementary" can carry several different meanings. In fact, science and theology are similar in some respects and dissimilar in others. In this chapter, we will consider their relation in some detail.

■ What Is Science?

The National Academy of Sciences explains science as follows:

> Science is a particular way of knowing about the world. In science, explanations are limited to those based on observations and experiments that can be substantiated by other scientists. Explanations that cannot be based on empirical evidence are not part of science. . . .Scientists can never be sure that a given explanation is complete and final. Some hypotheses advanced by scientists turn out to be incorrect when tested by further observations and experiments. Yet many scientific explanations have been so thoroughly tested and confirmed that they are held with great confidence.[1]

Science, therefore, depends on empirical evidence that can be verified by other scientists, and its theories are provisional, not final. There is, however, more to say about science.

First of all, most scientific hypotheses and theories are abstract and often contradict common sense. They are known by inference, not direct observation. We all know that the earth revolves around the sun, not the sun around the earth. How do we know this? If you were explaining this to a premodern person who believed what he saw—that the sun goes around the earth—what evidence would you provide? Has anyone ever actually seen the earth go around the sun? We believe this because we have been taught it on the authority of science, and have seen countless pictures of the solar system from grade school on. Yet no one has been in a space station above the solar system and observed the earth go around the sun. Astronomers infer that this is the case from observations and calculations of planetary positions. But they haven't actually seen it.

Consider another example. Recently news media have been reporting that astronomers are discovering new planets revolving around stars outside the solar system. Can they see these planets through telescopes? No. They have measured the light coming from distant stars, and noticed that there are tiny fluctuations in the light indicating that the star from which the light comes is moving toward us and away from us, to and fro, ever so slightly. They *infer* from this that the star's movement is caused by the revolution of a planet around the star. But they cannot see the planet.

Most people think that science is like common sense: it is seeing and touching things. But typically scientific theories, from relativity and quantum physics to evolution, are far from common sense. Relativity predicts that time slows down as one travels faster (though this is not apparent until one approaches the speed of light). So if a fifty-year-old man left earth on a space journey at nearly the speed of light, and returned forty years later, he would be younger on his return than his twenty-year-old son who had stayed behind on earth! The behavior of quantum particles is equally strange. Subatomic particles, like electrons, can be put in a sealed lead container. Yet there is a small chance that they will "tunnel" through the container and be found outside of it (so-called "quantum tunneling"). Physicists can predict mathematically what percentage of electrons will be found outside the container, but they cannot explain how it happens. Even simple scientific theories, like Galileo's law of falling bodies, require abstraction from common sense. Galileo hypothesized that a falling body accelerates constantly as it falls, and that a light body will accelerate just as fast as a heavy one, so that, in the absence of air resistance, a feather and a bowling ball would fall at the same rate. This certainly contradicts common experience. Galileo had to imagine bodies falling in a vacuum. He tested his theory by rolling balls down inclined planes (where air resistance would not be a factor). But neither Galileo nor anyone else can "see" the law of falling bodies. It must be inferred from evidence that we can measure.

Now there is a certain similarity here between science and theology. Theology also cannot verify its theories directly. You can't see God, or God's actions in the world. For this reason, many people distrust theological explanations, and place their faith in science. Science works, after all. But scientific theories cannot be seen either. They are inferred from what can be observed. And theology, in arguing for the existence of God, follows the same procedure as science, a process of inference from other things we believe to be true.

These same examples also indicate the biggest difference between science and theology. Science, while it can never be sure that its contemporary explanation is complete and final, can at least eliminate false theories with some confidence, by subjecting them to experimental testing. Consider the theory of the ether, referred to in chapter 5. For most of the nineteenth century, physicists thought that outer space was filled with the ether, a kind of gas that transmitted light waves (since waves could only be transmitted by some kind of medium). But experiments failed to detect the ether. Later, Einstein's general theory of relativity explained the transmission of light waves in another way. And so the theory of the ether was dropped. No one today defends it. Similarly, before Lavoisier discovered the true source of combustion, scientists thought that combustion was due to the release of a substance called "phlogiston." But no one now defends this theory, which has only historical interest. So the ether theory and the phlogiston theory are examples of false theories that, though they each commanded a consensus for over a century, were eventually discarded by science. By contrast, it is harder for theology to eliminate false theories, because they cannot be directly tested by empirical and laboratory procedures. (There is, however, a kind of lived verification of theology, which will be discussed later.)

Yet empirical testing is not the only criterion for the success of scientific theories, though it is an important one. Physicist George Ellis argues that there are four criteria by which scientific theories are evaluated. He lists these as: (1) simplicity, (2) beauty, (3) prediction and verifiability, and (4) overall explanatory power and unity of knowledge.[2] "Simplicity" means that scientists prefer the simplest theory that can accommodate the facts. "Beauty" may seem like an odd criterion. But many physicists, from Paul Dirac to Murray Gell-Mann, have insisted that a theory that exhibits mathematical elegance and beauty is probably true. Einstein's equation $E = mc^2$ is an example of a theory that explains a great deal, but which is both simple and mathematically elegant. "Prediction and verifiability" means that the theory is able to make predictions, often novel ones, which can be verified by observation and measurement. A striking example of this was the prediction of Einstein's theory of relativity that light waves would be bent in the vicinity of a heavy body, such as the sun. This was tested experimentally in 1919 when a team of astronomers measured the positions of stars during a solar eclipse, and found that the starlight was indeed "bent" to just the degree that Einstein's theory predicted. This phenomenon, which had been entirely unsuspected, was a striking confirmation of the theory. "Overall explanatory power and unity of knowledge" refers to the capacity of the theory to integrate and explain a large amount of data, and to fit in with the rest of contemporary knowledge.

Some of these criteria are more appropriate to one type of science than to others. Simplicity and beauty are particularly important in physics, but not so important in psychology. The ability of the theory to make testable predictions is also important in physics. But historical sciences, such as evolutionary theory, find it difficult to make predictions that can be verified empirically. Rather, they are judged primarily by other criteria, especially criterion four, broad explanatory power. Darwin did not make testable predictions in advancing his theory. But it explained and unified such a vast range of biological data that it eventually became accepted in the scientific community. This is exactly the same principle that is used in detective work. We cannot rewind the tape and replay the crime. We usually cannot see the crime, unless it has been videotaped. So we arrive at the best possible explanation, that is, the one that explains all the facts (broad explanatory power). Ellis comments on this:

The upshot is that in the case of the historical sciences, an element of interpretation is implied by any theory. . . . There is no way to avoid this. We minimize this as far as possible by demanding consistency with present day scientific theory and understanding, together with the requirement of broad explanatory power.[3]

Now, theology is similar to the historical sciences in this respect. It cannot easily make testable predictions. It is based on revelation in the past, which cannot be re-created for modern viewers. (We cannot rewind the tape to see Jesus performing miracles or to see the resurrection.) So it has to depend largely on other criteria: simplicity, beauty, and especially on overall explanatory power to validate its theories.

A second point to make about science is this. Science, as the National Academy of Sciences acknowledges, is "a particular way of knowing about the world." It is not the *only* way of knowing about the world. Our knowledge of our own consciousness and feelings is not scientific knowledge, yet it is the most immediate knowledge each of us possesses. We cannot prove scientifically that our friend or our spouse loves us, or that we love them. Likewise, our knowledge of God cannot be tested scientifically, nor can the conviction of many persons that there is a moral order that is independent of us and our opinions. In this regard, the British physicist Arthur Eddington once told an interesting parable. He compared science to a fisherman who fished the sea using a net with three-inch openings. Of course, he never brought up any fish smaller than three inches. After a while he concluded that there were no fish in the sea less than three inches long. This is like a scientist concluding that there are no nonmaterial realities because they cannot be detected by experiments. But of course that is a consequence of the method, not necessarily a feature of reality itself.[4]

Michael Polanyi has argued that some personal interpretation is built into every judgment, and that all knowledge, even scientific knowledge, therefore has a degree of interpretation built into it.[5] This follows logically from the admission, again acknowledged by the National Academy of Sciences, that scientific theories are not final and complete. They may be improved upon in the future; subsequent evidence or discoveries may reveal that the theories need revision. So they are interpretations, rather than perfect descriptions of reality.

Biologist Kenneth Miller argues that even in principle scientific knowledge cannot be complete. The reason is that quantum events are in principle unpredictable. Quantum theory can give us the probability of finding an electron here and not there, but it cannot make exact predictions about individual particles. It can only make statistical predictions about the behavior of a large number of particles. But a single unpredictable quantum event, such as the collision of a cosmic ray with a DNA molecule, could affect the course of evolution. So not only is the quantum world unpredictable, the macroscopic world is also. Here is Miller's explanation:

> Quantum physics tells us that absolute knowledge, complete understanding, a total grasp of universal reality, will *never* be ours. . . . It's not just a pair of colliding electrons that defy prediction. The mutations and genetic interactions that drive evolution are also unpredictable, even in principle.
> This is something biologists, almost universally, have not yet come to grips with. . . . The *true* materialism of life is bound up in a series of inherently unpredictable events that science, even in principle, can never master completely.[6]

So, even in principle, science cannot provide a complete explanation for the world. Events are, as the philosophers say, contingent, and to some extent unpredictable. John Polkinghorne's example (quoted in chapter 3) concerning the impossibility of predicting the movement of the air molecules in a room illustrates this. This example means that even everyday physical events are exquisitely sensitive to their context. Change the position of an electron 15 billion light years away, and it will affect the movement of the molecules in your room. There is no way that scientists or anyone else can know of the position of electrons on the other side of the universe. In other words, they cannot know what the ultimate context is. It could be God. And God could affect the state and position of quantum particles simply by the input of information, which would determine their indeterminacy.

So the claims that science will be able to explain everything are hollow. Even if physicists could reduce all the natural laws to a single formula, capable of being printed on a T-shirt, it would not explain the interaction of the air molecules in a room, or the course of events in a single human lifetime, or in a historical period, or in evolution. These are contingent and unpredictable. The dream of a deterministic, predictable universe, which we could control, died with the discovery of quantum indeterminacy and chaos theory.

For intellectuals (including scientists, philosophers, and theologians) there is a strong temptation to think that we might be able to explain everything with our theories. And if we can explain everything, we can control everything. But this is the sin of hubris, or, in theological terms, idolatry. We make idols of our theories by absolutizing them and not recognizing their limitations. Yet reality is more complex and mysterious than our theories allow, and it is wisdom to recognize this.

What Is Theology?

St. Anselm defined theology as "faith seeking understanding."[7] By this he meant that theology seeks to understand what believers accept on faith. If the Christian church teaches that God exists, Christian theologians try to understand God. If the church teaches that human beings are made in the image of God, theologians try to understand and explain what this means. To do this, theologians make use of reason, and of other kinds of knowledge, including scientific knowledge.

Theology is like science in one important respect: it claims to provide true knowledge, not just opinion. But the realm it deals with, God and God's relation to humanity and the world, is a different realm from that explored by science. This realm cannot be measured by instruments or tested by experiments. It is fundamentally not material, and therefore falls outside the scope of scientific method.

Theology has its own special methods and sources. It especially makes use of revelation and the ongoing experience of the worshiping community. Western theistic religions (Judaism, Christianity, Islam, Baha'i) and some Eastern religions, such as Hinduism, depend on revelation for their founding identity and Scriptures. Thus Judaism takes its beginnings from the revelations passed on through Moses, Christianity from Jesus, Islam from Muhammad, Hinduism from ancient sages, and so forth.

Revelation includes not only teachings (doctrines) but also a whole way of life. What was taught by Jesus was not just doctrines, but moral commands, attitudes such as love and dependence on God, and a manner of living. His miracles, which theologically are signs of God's coming kingdom, are also vehicles of revelation.

There is no exact parallel in the sciences to revelation, since revelation cannot be put to empirical test. There is no way we can re-create Moses, Jesus, or Muhammad, and then test to see if God was really speaking through them. Our knowledge of revelation is more like our knowledge of another person, something that cannot be measured by experiment. And it is like our knowledge of history, which, as we have seen, also cannot be empirically tested. Polkinghorne compares revelation to laboratory experiments.[8] By eliminating extraneous factors, experiments reveal an aspect of natural law more clearly than do ordinary events. Similarly, the lives of holy men and women reveal the nature and will of God more clearly than do the lives of ordinary people; they are, so to speak, windows onto God.

Revelation provides the foundation for religious communities and traditions. Usually this is set down in sacred Scriptures. But the ongoing experience of the religious community (including its leaders) provides the medium through which the revelation is interpreted and brought into the present. Many Christian beliefs and practices owe their existence not simply to Scripture, but to the experience of the worshiping community. It is not enjoined in Scripture that Sunday should be the day of worship, or that infants should be baptized, or that marriage should be celebrated as a Christian ceremony. Most theologians do not believe that the nature of the Trinity or of Jesus (fully God and fully human) is spelled out in Scripture. Rather, these foundational doctrines were worked out through hundreds of years of community experience and reflection: faith seeking understanding. On the other hand, there are scriptural injunctions that are no longer observed. The decree of the apostles in Acts 15 directs the community to require that Gentile converts should abstain from "things polluted by idols and from fornication and from whatever has been strangled and from blood" (Acts 15:20). Yet this provision was gradually dropped from Christian practice, even by very conservative Christians.

There is therefore a kind of development, both of understanding and of practice, within Christianity (and other religious traditions). No Christian theologian or community today would support the practice of slavery in any form. Yet it seems to be endorsed, or at least condoned, by the New Testament. Ephesians 6:5 reads "Slaves, obey your earthly masters with fear and trembling . . . as you obey Christ." Similarly, 1 Timothy 6:1 reads "Let all who are under the yoke of slavery regard their masters as worthy of all honor." The realization that slavery is against the will of God, and is never justifiable, only gradually developed within the Christian community. Now it is recognized that if each person is an image of God, and therefore has inalienable rights, slavery is simply incompatible with the Christian gospel. There is therefore a gradual development in the Christian understanding of the meaning of Christ's revelation. Vatican Council II, in its *Constitution on Divine Revelation*,[9] acknowledges this development:

> This tradition which comes from the apostles develops in the Church with the help of the Holy Spirit. For there is a growth in the understanding of the realities and the words which have been handed down. This happens through the contemplation and study made by believers . . . and through the preaching of those who have received through Episcopal succession the sure gift of truth. For, as the centuries succeed one another, the Church constantly moves forward toward the fullness of divine truth until the words of God reach their complete fulfillment in her.

Thus there is a development of theological understanding. In this also theology is similar to science.

■ Science and Theology

Science and theology, then, are similar in some respects. Both claim to provide real knowledge about the world. Both claim that there is a development in their knowledge, that is, that their theories about the world become increasingly more accurate descriptions of reality. Both would recognize that there is an element of human interpretation in their theories. That is, the theories do not perfectly describe nature, as the National Academy of Sciences recognizes (in the quote given at the beginning of this chapter). If they did, there would be no more possibility for increasing knowledge.

The way scientists' language reflects this recognition is that they often talk about models. A model is not a scale model, like a model ship or airplane; it is an abstract representation, often mathematical, of the reality scientists are attempting to describe and predict. So physicists might speak of the Bohr model of the atom, or a computerized model of a weather system, or a model of how evolution works. Models are related to hypotheses and theories, which are more complete explanations of what the model seeks to represent in simplified form. What is important about models is that everyone recognizes that they incorporate an aspect of human interpretation. They are not perfect maps of reality. Like most maps, they are approximations, which are more or less accurate.

Theologians also use models. They might speak of a "model" of God, for example. What would this mean? It would mean that we think of God as having certain characteristics. The Bible portrays God as a loving parent, but also as a king, as cosmic Creator, as loving but also as capable of anger, as just, but also as merciful, etc. During the Middle Ages, people thought of God as a powerful king. This model is not so attractive to modern democratic Christians, who prefer other models, especially God as loving father, or perhaps mother.

Both scientists and theologians would claim that some models are better than others. "Better" here means that they correspond better to the reality they seek to describe than some other models. (This is known in philosophy as the correspondence theory of truth.) Science, as we have seen, is often able to eliminate poor models (and poor theories) by empirical testing. This is the great strength of the scientific method. For if one can eliminate false models, one can come to some general agreement on what the true models might look like. One of the most famous philosophers of science, Karl Popper, argued that this is all science can ever do. It can disprove false theories, but it cannot prove that theories are true, because there is always a chance that some further evidence will show that a given theory is inadequate. Many people, however, think that Popper understated the cumulative power of the science method. There seem to be some scientific models and theories, such as the atomic theory of matter, that are so well established that they are not going to disappear. Yet it is possible that some later theory will modify the present atomic theory. String theory, for example, argues that the fundamental units of matter are tiny, vibrating strings. And experiments indicating nonlocality may show that individual atoms are in some mysterious way connected with other atoms distant from them. So the atomic theory of matter, while a good approximation, may not tell us all there is to know about the nature of matter.

The same is true of theological models. We can argue that they are good approximations, but not that they tell us the whole truth. We say, for example, that God is love (1 John 4:16). This is certainly a better model than "God is hate." But it is not a complete model either. It is a simplification, which characterizes God in terms of a limited human quality. God's love is not exactly like human love. There are many kinds of human love: a love of desire (*eros* in Greek), a love of friendship (*philia* in Greek), or a self-giving charitable love (*agape* in Greek). Even if we specify God's love as *agape*, it still is not just like human *agape*. For human love is limited and finite; God's love is unlimited and infinite. Furthermore, human love exists in time, and therefore changes, but God and God's love exist in eternity and do not change. So, to characterize God as love is a good approximation, from our point of view. Yet it incorporates elements of human interpretation and limitation. We cannot claim that it is the whole truth about God. Similarly, we speak of God as "Father," Jesus' own term. But no one thinks that God is just like a human father: that he has a body, a specific location in space, that he will get old and die, etc. God is like a father in one or two respects: He creates us, as our father (and mother) did, and he loves us like a loving father, but to an infinitely greater extent.

I have said that the great strength of science is that it can eliminate false models by empirical testing. This is not quite as unambiguous as it might seem. In the last analysis the final court of appeal must be the relevant scientific community itself. It is harder for theology to eliminate false models, precisely because the reality it investigates, spiritual reality, cannot be empirically tested, at least not directly. And yet theology does progress. Some practices, like slavery, can be seen in retrospect to be incompatible with Christianity. How, then, does theology eliminate false beliefs and practices? I think there is in the life of the worshiping community a kind of lived verification. If the belief or practice proves to be destructive of Christian values in the long run, it will cease to recommend itself to practicing Christians. This is what happened with slavery. Jesus himself refers to this kind of testing. When he is asked how to tell true prophets from false prophets, he answers: "You will know them by their fruits" (Matt. 7:16). Presumably "fruits" here means the fruits that come from a holy life. St. Paul gives us a list in Galatians 5:22–23: "love, joy, peace, patience, kindness, generosity, faithfulness, gentleness, and self-control." But just as we can recognize good or bad fruits in the work of a prophet, so can we recognize them in the life of a community. Certain practices, like slavery or witch burning, bear bad fruit. But some doctrines do also. The traditional Calvinist doctrine that all persons are predestined to heaven or to hell has not commended itself to most Christians, and has been largely repudiated even by most of those in the Calvinist tradition.

Conversely, some Christian doctrines can be affirmed through the lived experience of the community. An example of this is the doctrine of grace. "Grace" means the aid or help that God gives to strengthen and elevate the will of believers. Classic texts on this are found in Paul's letters and in the *Confessions* of St. Augustine. A lived verification of this teaching is found in the experience of Alcoholic Anonymous and twelve-step programs generally. These programs have been enormously successful in helping addicts to escape addiction by turning the problem over to God (or a higher power) and relying on God's help. Their programs are simply a lived application of Paul's teaching on grace, and their success has helped to confirm the truth of that teaching on grace.

Other Christian models and theories are harder to verify. Take the model of God as Trinity. This competes with the Jewish and Islamic model of God as nontrinitarian. Can one determine which model is correct by a process of lived verification? Some might claim that the lives of Christians are clearly much more holy, much more in accord with the fruits of the Spirit, than are those of Jews and Muslims, and that this would therefore confirm the Christian theory of God. But many would deny this, so there is no consensus.

I agree with John Polkinghorne that this is because the object of theological investigation, God and God's relation with the world, transcends us, whereas in the case of the natural sciences, we transcend the object being investigated. This means that we can put nature to the test, but we cannot put God to the test.[10] And so it is less easy to verify or falsify theological conclusions than those of natural science. This is one of the great differences between science and theology, and it stems from the fact that the two disciplines investigate different aspects of realty, and so must use different methods. The scientific method of empirical testing is simply not adequate for an investigation of God.

■ Ways of Relating Science and Theology

There are several different ways of relating science to theology. I will follow here the typology proposed by Ian Barbour, because it is the most widely used. Other authors, however, such as John Polkinghorne[11] and John Haught,[12] have proposed slightly different typologies.

Barbour proposes a four-fold typology of the relation between science and religion: conflict, independence, dialogue, and integration.[13] Conflict is self-explanatory. Science and religion conceive of themselves as in conflict. This is the position characteristic of hard naturalism, as I have described it here. Nature is all that exists; science can explain everything and is the only legitimate form of knowledge. Religion and theology are fictions, outmoded and antiquated modes of thinking that have been displaced by the sciences. Richard Dawkins, E. O. Wilson, P. W. Atkins, Francis Crick, and others are representatives of this group. Barbour also classifies some fundamentalist Christians in the "conflict" group, especially those who reject evolution. For they, in parallel with the naturalists, think that to accept science they must abandon their religion. Hence there is a conflict. If science is true, religion is false. You can choose one side but not both.

The second way is independence. Practitioners of this model of thinking imagine that science and religion are two completely distinct realms of inquiry and modes of thought, neither of which can have any influence on the other. Science and religion might be seen as separated by differing languages or differing methods, or, most commonly, as investigating entirely distinct realms of reality: the material and the spiritual. A good example of this is the late Stephen Gould's NOMA: that is, "non-overlapping magisteria." Gould thought that science and religion each had their own competencies in different areas. As he wrote: "[T]he net, or magisterium, of science covers the empirical realm: what the universe is made of (fact) and why it works this way (theory). The magisterium of religion extends over questions of ultimate meaning and moral value. These two magisteria do not overlap."[14]

Many scientists who are also believers espouse this position. Frequently they will say that science deals with fact and knowledge, religion deals with faith and values. Or: science deals with knowledge and matters of fact, religion with the heart. Independence is a simple way of keeping these two areas separate, and therefore of preserving religious belief from the corrosive acids of scientific skepticism. I suspect that many who hold this position do so because they are not sure that their religious beliefs would survive scientific scrutiny. If they were forced to bring their religion and their science together, they might lose their faith.

Dialogue includes a diverse group of positions that hold that science and religion are distinct, but that also hold that they can influence one another. Some, like John Polkinghorne, think that science raises questions that science itself cannot answer, but which can be answered by theology.[15] Others hold that there are methodological parallels between science and theology: both make use of models, both seek knowledge, both must rely on personal judgment in the act of knowing (Michael Polanyi). Finally, various forms of nature-centered spirituality can be included under the "dialogue" position.

The "integration" position attempts to integrate the findings of both science and theology. Barbour gives three examples of this: natural theology, a theology of nature, and systematic synthesis. Natural theology attempts to argue from the perception of design in nature to the existence of God. A theology of nature considers the theological implications of some particular scientific theory, such as evolution. A systematic synthesis attempts to integrate both science and theology within a more comprehensive philosophy. Examples of this are Aristotelian and Thomistic philosophy up to the seventeenth century, emergentism (discussed above), and process philosophy.

I will defend here the complementarity, which is a species of dialogue, of science and theology. Both the book of nature and the book of Scripture reveal God, and we need both to get a comprehensive account of reality. Science and theology are not, however, complementary in the sense of being independent from one another. It is true that each explores different domains: science the physical and theology the spiritual. And it is true that the methods of science do not lend themselves to discovering spiritual truths, since these cannot be tested empirically nor quantified mathematically. So there is a distinction between the domains searched by science and that explored by theology. But at the same time, if spirit and matter are related, and if God does act in the world, then these domains intersect in some areas. The most obvious is the nature of the human person. We can give a scientific account of the physical processes going on in a person, but it is very hard to give a coherent scientific account of free will, mental causation, spiritual and mystical experiences, afterlife, and so on. Science also cannot discern purpose or meaning of human life, or any moral values that are more than subjective or cultural conventions. For all of this we need philosophy and theology. But the human person is not split into two different realms: one physical, the other spiritual. So we need science and theology in dialogue to reach a complete picture.

A familiar example of this is the treatment of alcoholism. The professional literature usually analyzes alcoholism as due to physical and genetic influences, as well as environmental and cultural influences.

It seldom mentions free will or spiritual factors. Yet anyone familiar with alcoholism knows that the free will—the decisions made by the alcoholic—play an enormous role in recovery. It is not as if the alcoholic has no choice. But this is the conclusion toward which the modern climate is leading. Finally, the various twelve-step programs have proved to be very helpful in aiding alcoholics (and other addicts) to overcome their problem, often when no other treatment worked. For example, Roger Johnson, a psychiatrist practicing in Roseville, Minnesota, tells the following story. In 1988, he was in treatment for alcoholism at St. Mary's Hospital. He was, in his words, "at the end of my rope." He had been an agnostic, but began praying for his condition. He asked God to take away his urge to drink. The next day, his desire to drink left him, and he has been sober ever since. Many such stories can be found in the so-called "Big Book" published by Alcoholics Anonymous.[16] So spiritual healing also should not be ignored in the treatment of alcoholism. But because it and free will do not fit easily into the naturalistic and scientific paradigm (they cannot be quantified or easily tested experimentally), they are generally ignored in the professional literature concerning alcoholism. Thus, to get an accurate model of alcoholism, we need to consider four factors, not two: genetics, environment and upbringing, free will, and spiritual factors. But these same four factors affect almost all human behavior, and therefore need to be considered in any adequate theory of the human person. In other words, we need both a scientific perspective and a theological perspective.

In previous chapters, we have seen that bare scientific accounts of the Big Bang, of evolution, and of the human person, when left to themselves, cannot provide any satisfactory account of ultimate purpose or meaning, of moral values, of free will, of mental causation, or of the human experience of God. Usually what happens is that some species of naturalism fills this gap. We end up being told that there is no God, (Dawkins, et al.), that the only way of knowing is scientific knowing (E. O. Wilson), that the universe created itself (P. W. Atkins), that evolution is a blind, random process (Dawkins), that humanity is an accidental, meaningless spin-off of this blind process (Gould), that human beings have no free will, no hope of afterlife or ultimate consolation, and so on. But if we assume that God exists and created the universe and humanity, we have a picture that is a more accurate representation of our total experience as human beings. Thus science and theology are complementary: each completes the account of the other. Science, because of the limitations of its method, is a powerful way of knowing some features of reality, but not all of reality. Theology, which uses a somewhat different method (though there are similarities to scientific method), helps account for other, more spiritual, aspects of reality.

■ Process Theology

In an important book, *Religion and Scientific Naturalism,* David Ray Griffin has argued for what he calls a "minimalist naturalism," which should be acceptable to science but also provides substantial support for religious views. He defines "minimalist naturalism" as follows. "In the *minimal* sense, scientific naturalism is simply a rejection of supernatural interventions in the world, meaning interventions that interrupt the world's most fundamental patterns of causal relations." Griffin goes on to say that science need not be committed to any more restrictive dogma. It can be open, for example, to the possibility of genuine purpose, freedom, or even life after death. "Scientific naturalism could, therefore, be compatible with a naturalistic theism or theistic naturalism."[17] But what Griffin means by "naturalistic theism" is actually process theology.

We cannot go into the details of process theology here. It has many followers among scientifically minded theists. But it differs from traditional Christianity at a number of points, especially in its conception of God. Process thought is based on the philosophy of Alfred North Whitehead. Whitehead rejected the "monarchial" model of God, in favor of a God who loves and influences nature and human life by persuasion. God is not omnipotent, did not create the world from nothing, and cannot interfere in natural processes, but he is an omnipresent "lure" attracting each minuscule event in the direction of novelty, beauty, love. Whitehead writes: God "does not create the world, he saves it, or, more accurately, he is the poet of the world, with tender patience leading it by his vision of truth, beauty, and goodness."[18] But even God cannot interrupt the web of finite causation. Therefore Whitehead's God cannot intervene to perform miracles. Furthermore, God develops with the development of the universe. In God's "primordial" state, God is "the unlimited conceptual realization of the absolute wealth of potentiality. . . . [W]e must ascribe to him neither fullness of feeling nor consciousness."[19] God's consequent state is the result of the development of God and the universe together: "[T]here is a reaction of the world on God. The completion of God's nature into a fullness of feeling is derived from the objectification of the world in God."[20]

Process thought has significantly influenced modern conceptions of God. Process thinkers have criticized the classical idea of Aquinas and others that while the world is dependent on God, God is not dependent on the world. Aquinas thought that God, as eternal, must be unchanging. Therefore God could not empathize with or suffer with his suffering creatures (for to do so God must change). Thus God seemed to be detached and remote from the world, unlike the biblical portrait of God, who loves his people as a mother loves her nursing child (Isa. 49:15). Many theologians now will hold that God is indeed related to the world, and suffers with it. W. Norris Clarke has proposed that Aquinas's idea of God as the fullness of actuality needs to be complemented by allowing for an element of receptivity in God.[21] Others would hold that God suffers with the world through Jesus, who retains his human nature in the resurrection, and remains related to us as our fellow sufferer.

While process philosophy has made contributions to theology,[22] it has met with several criticisms. First, it asserts that processes in the world are made up of tiny episodes—time is really a succession of atomic events—rather than being continuous. But scientists point out that electrons do not exist only episodically, as a succession of events. Rather, electrons and other particles seem to exist continuously, as discrete forms of energy with specific masses, charges, etc. John Polkinghorne, a particle physicist and a theologian, explains: "Quantum physics involves both continuous development . . . and occasional sharp discontinuities (measurements) but it does not . . . suggest the 'graininess' that process thinking seems to suppose."[23] This same criticism would apply to the self. Is the human self really only a chain of episodes? What then accounts for its continuity? Finally, many cannot accept the God of process thought. Such a God cannot create the world from nothing nor intervene in its processes to bring about miracles; he can only gently persuade. Such a God seems far more limited and powerless than the God of Jewish and Christian theism. Furthermore, the God of Jesus is not merely a God of persuasion; he is also a God of judgment, as many New Testament passages testify (Matt. 13:41–43, 47–50; 25:31–46; etc.)

Thus, although I am sympathetic to Griffin's argument in many respects, I cannot follow him in his acceptance of process theology. He is explicit in rejecting God's creation of the universe from nothing, and the possibility of miracles, or any kind of divine "intervention" beyond God's omnipresent "lure" or persuasion. Though Griffin wants to hold for the possibility of life after death, it is not clear how this is possible within the limits of his system.

It is also not clear why science has to assume that there can never be events that transcend natural causality or what we know of natural causality. How could scientists possibly know this? Have they surveyed every event in all the billions of galaxies in the universe so as to satisfy themselves that all of these events can be fully explained by natural and material causality? No conceivable experiment could show that the web of natural causality is never transcended. As we have seen, we cannot even account fully for the nature of matter. Its roots may extend into higher dimensions. How can we possibility assume then that we understand even the limits of natural causality? To be fair to Griffin, he does believe in the reality of paranormal events, including extrasensory perception. So he is willing to affirm that what we call "natural" causality extends farther than what the present scientific consensus allows. But why not then allow for the possibility of miraculous causality as well? It is not true that one must dogmatically affirm that only natural causality exists in order to be a good scientist. Peter Hodgson, for example, a physicist at Oxford who has published hundreds of scientific papers and several books, and who is a frequent Templeton lecturer, is a Catholic who believes in miracles.[24] But this has not inhibited his scientific work. It seems to me that all a scientist needs is the working hypothesis that probably the phenomenon under investigation can be explained by natural causes, and that science can therefore explain these causes. If it eventually is the case that science cannot explain a particular phenomenon (such as, for example, nonlocal interactions), then that may mark a limit to scientific knowledge. Or it may be that later generations of scientists would be able to explain the phenomenon. But there is no reason why scientists must embrace the unprovable dogma that natural causality is never transcended at any time, anywhere in the universe.

■ Why Science Needs Theology

Science can help us understand a great deal about God. The creation itself manifests the glory of God (Romans 1). Science has revealed to us the grandeur and mystery of the universe, its seemingly infinite extent, its marvelous subtlety and intricacy, its incredible diversity, the simplicity of its fundamental laws. All of this reflects its Creator. But science, left to itself, usually concludes in a form of Deism. To understand God's personal and loving qualities, and God's transcendence, we need biblical revelation. This we cannot get from the study of nature itself, for nature does not seem to care for individuals or even for species (most of those which have existed are now extinct). And nature cannot transcend itself. In spite of its energy, power, and beauty, it ends in death, for individuals, species, planets, solar systems, stars, and galaxies. All of its beauty, whether that of the mayfly or that of the star, is transient. And so, as St. Augustine realized, it points beyond itself to that which is forever beautiful, forever young, forever alive. But to know this source we need revelation and theology.

Does theology have anything to contribute to science? Most scientists would probably say, "No." One can do excellent science without knowing anything about theology. Indeed, a prevalent view is that religion and theology retard scientific investigation. Science advances by pushing back religious obscurantism and superstition. This has been a persistent theme among the apologists of modern science, from August Comte and Sigmund Freud to contemporary authors like Richard Dawkins and E. O. Wilson.

I will argue the opposite. Modern science came to birth in postmedieval Western civilization. This was a civilization founded on the belief in God as a rational Creator and sustainer of the universe, a belief shared by Jews, Muslims, and Christians. A number of authors have argued convincingly that the theistic belief that the universe was designed and sustained by an omnipresent rational Creator was crucial for the rise of modern science. Alfred North Whitehead, for example, argued that "the medieval insistence on the rationality of God" led to "the inexpugnable belief that every detailed occurrence can be correlated with its antecedents in a perfectly definite manner, exemplifying general principles."[25] And this belief in a universal natural order in turn formed the foundation for modern science. Therefore, according to Whitehead, "faith in the possibility of science . . . is an unconscious derivative from medieval theology,"[26] for science must assume that the universe, in its ordinary operations, is everywhere governed by constant, rationally knowable laws. The whole idea of natural and physical law derives from the belief in a rational lawgiver, namely God. That is why the term "law" is used to describe scientific generalizations (which actually, in their present form, are not like legislated laws, but are usually couched in the language of mathematics). Furthermore, it was crucial to modern science that the order of the universe be understood as contingent, not as necessary.[27] The mistake made by the ancient Greeks was to assume that the order of the universe was necessary, and could therefore be deduced from first principles by reason alone, without experimentation. But if the universe is contingent, then it could embody any number of rational structures, and experiment is necessary to discover just what order actually does exist in the universe. Thus it was the combination of a belief in a universe ordered by rational laws and a belief in a contingent universe, which had to be investigated experimentally, that gave birth to contemporary natural science.

Now, one can hold, as Ian Barbour does, that to begin with, science needed the context of a rational Creator who ordered the universe according to universal, intelligible laws. But now, science is self-sustaining, and can dispense with the religious and theological foundations on which it was built.[28] I question this. It seems to me that the loss of any religious support for a universal, objective rationality to the universe will result in the universe being perceived as random and chaotic. Also, it is likely that if there is no foundation for an objective rationality and intelligibility, people will conclude that one opinion is as good as another. This is already the common belief among college students. It is powerfully reinforced by postmodern philosophies, which tend to explain all beliefs as due to sociological or biological factors. Scientific naturalism plays right into this current of thought. For, as we noted earlier, if my thoughts are determined by my neural networks, which are ultimately programmed by my genes and environment, and the same is true of you and everyone else, then there is no reason to conclude that my opinion, or yours, is true. Rather, each of our opinions is no more than the product of our genetics and our upbringing. There is no mental causation—one idea does not lead to another—and we cannot freely choose between alternatives anyway. So our thoughts are determined, in the same way that computers are programmed. Francis Crick virtually admits this when he says that free will is an illusion, and that our decisions are determined by unconscious processes going on at the neural level. There is therefore no objective truth, or if there is, we cannot know it. In such a world, it would not be truth that persuades people; it would be force that decides which opinion and which point of view will prevail. There can be no other arbiter.

Thus, in the end, science itself is undone by scientific naturalism, a surprising irony, but, it seems to me, inescapable. The loss of a belief in God will undermine any widespread belief in any universal, rationally knowable order, and precipitate us into a world of chance and subjectivity, in which one person's opinion is as good as another's. In such a world, science cannot flourish. And therefore, science benefits from the religious and theological belief that there is a rational Creator and sustainer of the universe, who also sustains the universal laws by which the universe operates. In fact, both science and (classical) theology share a deep commitment to a rationally knowable and intelligible order that underlies the world. In this respect they are like sister disciplines working in different fields.

10

Conclusion
Beyond Naturalism

■ Chapter 1 lays out four integrating themes that run through this book. Let us reconsider these themes in the light of what has been discussed in the preceding chapters. The first theme was the challenge of naturalism and the historical process that led up to it. The second was that modern naturalism is the result of our view of nature as a self-sufficient system, which seems to have no need of God. In the industrialized world, God is fading away from consciousness like the Cheshire cat in *Alice in Wonderland*. For many people, all that remains is the grin. The third theme is that Christianity actually can give a better rational account of the world and humanity than naturalism can. And the fourth is that Christianity and science need to be partners. Science has much to contribute to Christianity, but conversely, Christianity has much to give to science. Let us then reflect further on each of these themes.

With respect to the first theme, naturalism is clearly winning the mind game in the academy—the universities and research centers of the West (and more and more of the East). British philosopher of religion John Hick puts it this way: "Naturalism has created the 'consensus reality' of our culture. It has become so ingrained that we no longer see it, but see everything else through it."[1] But it would be a mistake to assume that naturalism can be confined to the universities, where it only affects a handful of intellectuals. For the universities educate the bulk of the population, including secondary school and elementary school teachers, journalists, television personalities, filmmakers; in short, the shapers of public opinion. So, although there is a cultural lag, once an ideology has been established within the walls of the academies, it will inevitably diffuse out into the general population. Again, one could reply that most of the population now lives in nonindustrialized countries, where modern agnosticism has not yet gained hegemony. But nonindustrialized countries aspire to become like the industrialized countries. Once the elite of traditional countries, for example, India, become educated in the sciences, they tend to lose their traditional religious faith.

Of course there is a reaction to this trend in the West, especially in the United States. Many people take refuge in so-called fundamentalist Christian groups, which reject a good deal of modern science, especially the theory of evolution. Other people join so-called "New Age" groups, whose religion draws from Eastern and Native American religions, but which also embraces some superstition. But neither of these strategies is a good solution to the problem of naturalism. Like it or not, science is with us to stay. The solution is not to reject it, or to retreat into premodern superstitions, but to redeem it. For, if I am right, many scientists have made an idol of nature. Nature is their only reality, their ultimate context, and the object of their hopes for salvation. Much of the premise of contemporary scientific industrialism is that we can save ourselves by mastering nature: better technology, better drugs, and so on. But even a glance at the history of technology is enough to show that this is an illusion. Popular magazines in the fifties were full of the promises of atomic energy for unlimited energy. Not many foresaw the arms race, the problems of radioactive waste, the threat of nuclear annihilation, and so forth. Similarly we were regaled with articles describing television as a great boon to education. But no teacher I know thinks that today's young people are better educated than those of a generation ago. Students know little about English, history, science, and even geography, something that is primarily visual and that should be taught well by television. In short, television has been a disaster for education. The point is that every technological advance is a two-edged sword: some good comes out of it, but also some evil. Salvation through technology is therefore an illusion. This should not be a surprise to Christians, for our religion has held from the beginning that we cannot save ourselves.

The second theme is that God now seems remote from us. We have traced the reasons for this in chapters 2 and 3. Modern science, beginning in the seventeenth century, has increasingly come to see the natural world as a self-sufficient system that needs no greater context to support it. It is no longer a sacred cosmos that communicates God's presence. Because God seems so remote, many people may have a vague belief in God, but that belief does not seem to affect their lives either at work or at play. This amounts to what I have called practical naturalism: people act as if the natural, physical world is all we can be really certain of, all that we can really put our trust in. In this regard, it is worth pointing out that the Greek word for "believe in," *pisteuein*, carries a broader and deeper range of meaning than does the English word "belief." To say *pisteuo* in God is to say that I believe God exists, I commit myself in trust to God, and I rely on God. *Pisteuein* indicates a much deeper level of commitment than "belief," and this depth of commitment is what the New Testament means by "faith."

The superficiality of much contemporary belief in God is reflected in consumerism and a hedonistic lifestyle—the "live for today, live to have fun, live for distraction and entertainment" attitude. But another indication is the widespread depression and pessimism characteristic of Europe and increasingly of America, for the problem with naturalism is that we are confined to our natural situation. For the elite, such as professors at major universities, this may not be bad news. But it is very bad news for Joe and Jane Sixpack, or for the millions trapped in the cycle of poverty, malnutrition, low-paying jobs or joblessness, broken families, disease, alcoholism, and despair. Most of these people have only average or below average abilities and are poorly educated. For them, there is scant hope without God. This widespread hopelessness is evident in the media. Most contemporary films are either sentimental and escapist (e.g., Disney films) or are brutal, pessimistic, lonely, and tragic (especially European films). Contrast this with Chaucer's *Canterbury Tales*. Chaucer wrote these tales between 1385 and 1399, only about fifty years after the worst tragedy in European history, the Black Death, which, between 1347 and 1351, carried off one-third of the European population. Moreover, the plague returned at intervals until the 1600s. Yet Chaucer's tales, though ribald and baudy, are full of a *joie de vivre* and an optimism that is rare in modern film or fiction. Behind Chaucer's pageant is the sense that, in the end, "all will be well," to borrow the words of Julian of Norwich, a fourteenth-century English mystic. And this optimism came from his Christian faith. But how many of our contemporaries have the deep certitude that, in the end, "all will be well"? More and more of us find our consolation in antidepressants, alcohol, drugs, or mindless distractions.

The third theme is that a Christian viewpoint can explain the facts of humanity and of modern science better than naturalism. Chapters 4 through 7 of this book endeavor to sketch out such an explanation. I think this is particularly true when we include (as we must) human beings in the panoply of what is to be explained. I have focused above on free will and mental causation—the fact that we can freely reason from premises to conclusions without our thoughts being determined—as significant. But there are other features of human life that also are not explained well by naturalism. Most human beings, worldwide, believe that some deeds are right and some are really wrong. Not many persons would agree that our cultural laws against rape, against brutalizing and torturing children, and against murder are merely genetically driven or culturally conditioned rules, like driving on the right side or the left side of the highway. John Polkinghorne put it well when he once said (in conversation) that he knew that torturing infants was wrong, and that he knew this as certainly as he knew anything. But according to a naturalist outlook, all moral codes are merely the products of genetics or of culture. There can be no justification for a universal moral code, because there is no universal lawgiver or universal moral law structured in nature. Therefore (though this is a controversial point) I cannot see how, in the absence of God, one can maintain that anything is "really" or ultimately right or wrong. I might think that torturing infants is really wrong, but how could I argue that my belief should be binding on others unless there is one God who created us all? And indeed, this was historically the reason given not only for universal moral codes but also for universal human rights.

According to the genetic theory of altruism, popularized by Richard Dawkins in *The Selfish Gene* and learned by most college students in biology, human altruism develops as a way of ensuring the survival of one's genes. If I sacrifice my life for two or three of my brothers, I am ensuring the survival of my genes, since my brothers share half my genes. Now this, indeed, does explain what might be called "Mafioso morality." In the Mafia, as in other clan-based moralities, one defends one's kin and attacks strangers. This morality is widespread and might be called the "default" morality of human culture. But what is striking is that it was precisely this kind of kin-based morality that most of the great religions of the world have opposed. The great prophets of Judaism announced that God was indeed the God of all persons. God cares for even the enemies of Judaism, such as the Assyrians (cf. the book of Jonah), who therefore must be treated morally. Jesus said that he who loved his father or mother more than him was not worthy of him (Matt. 10:37). In Christianity, the measure of discipleship is love, especially of enemies and strangers (cf. the parable of the Good Samaritan—Luke 10:29–37). Likewise, Muhammad claimed that all Muslims of whatever tribe were brothers and should be treated as such. The recognition of universal moral rules that go beyond clan morality is also prominent in Eastern religions, such as Buddhism. This raises the obvious question: If morality is genetically driven, and is like the rules governing wolf packs (as one biology student put it to me), why would the universal moralities found in the great religions even arise at all? Why wouldn't all morality, in all human cultures, be clan or Mafia morality? Why is it that some people do indeed sacrifice their lives for strangers, either in a single moment, to prevent catastrophe, or gradually, over a lifetime of service, as Mother Teresa's Sisters of Charity do? For these questions, naturalism has no answers.

Another prominent feature of human life is the nearly universal human need for religion, which goes back to at least Cro-Magnon times. Naturalists generally explain this by arguing that religion gives human groups an evolutionary advantage because it enforces bonding among the group. There are real problems with this explanation. First of all, traditional cultures commit a large amount of time and energy to religion. Recall the Cro-Magnon burials in Sungir, Russia, 28,000 years ago, in which three persons were found, each dressed in clothing onto which more than three thousand ivory beads had been sewn. It was estimated that each bead took about one hour to make.[2] This is a huge sacrifice of labor and materials to make for the sake of religion. Why, in the face of scarce resources, wouldn't human groups develop that invested little or none of their resources in religion, but invested instead in making better weapons, perfecting their hunting skills, and so on, and at the same time solving the problem of human bonding by simply developing an instinct for group loyalty? Indeed, do we not observe such deep-seated group loyalty even today, unleashed in the name of nationalism, and even in loyalty to sports teams? Such groups would have a huge advantage over those groups that sacrificed their resources for religion, and should have driven them from the field through natural selection, because a more efficient use of resources would translate into better survival of progeny and hence into evolutionary success. Indeed, if we wanted to turn the naturalist argument on its head, it would be tempting to argue that perhaps the reason that religion is so widespread in human groups is because it actually works! Perhaps coming into contact with God does help us after all, as Deuteronomy 28 and the Psalms promise, even in such mundane challenges as survival.

In any contest between theories, it is the theory that best explains ALL the facts that is the best explanation. We use this routinely in everyday life. The explanation that explains all the problems with a malfunctioning automobile engine is better than the theory that only explains some of the problems. The theory that explains all the facts of a murder case is better than one that explains only some of them. The same is true when we are trying to explain the universe and human life. A Christian explanation, such as I have developed in this book, can incorporate the findings of contemporary science, but can explain in addition other facts, such as the human experience of free will, and of the freedom to construct chains of reasoning that are not determined by our neurons, the widespread human need for religion and sense that there are universal moral rights and wrongs, and finally the belief, found in all cultures, that there is a part of the human person that survives death in an afterlife. Christian theology can explain these features of human life, but can also help explain scientifically derived features of the universe, such as the Big Bang and the anthropic principle (chapter 4). Nor is such reasoning only appealed to by theologians. Physicists George Ellis (*Before the Beginning*) and John Polkinghorne (*Serious Talk, The Faith of a Physicist*) argue in just this way in their defenses of Christian belief.[3]

It is true that Christian faith cannot be proven by reason. But this does not mean that reason cannot support faith. The first letter of Peter exhorts us: "Always be ready to make your defense to anyone who demands from you an accounting for the hope that is in you" (1 Peter 3:15). Christian faith was never meant to be blind faith. For, if it is, the Christian cannot answer the skeptic's question: "Am I to believe in every absurdity? If not, why just in this one?" Faith goes beyond reason, certainly: we can never prove that God is triune, or that God loves us personally. But we can develop a theology—an understanding of Christian faith—that shows that faith is consonant with reason, including scientific reason. And in the end, this is a stronger faith than a blind faith that can marshal no reasons in its defense. It is also true that faith is a gift of the Holy Spirit (1 Corinthians 13); we cannot argue our way into faith. But a reasonable theology can at least clear away many obstacles that might prevent even well-disposed persons from coming to faith.

The fourth theme that I have stressed is that theology and science should work together as partners, not as enemies. My reasons for this are given in chapter 9. Christianity needs science to help purge it of superstition. Science can also increase our appreciation for the beauty and awesomeness of the natural world. And, of course, it grants many practical benefits to human beings, such as medical treatments for disease. But Christianity, if I am right, also has gifts for the sciences. It underwrites the conviction that the universe, because it is the creation of a rational Creator, is intelligible and not chaotic and unknowable. It underwrites the conviction that human beings have freedom of choice and the freedom to construct rational arguments that are not biologically determined. Without this, science could not claim to be objective; it could only claim to be one perspective among many. Christianity underwrites values such as honesty, selflessness, the importance of objectivity, and the importance of knowledge, which are necessary for the practice of science itself. And finally, if I am right, it has the capacity to release scientists from their captivity to the ideology of naturalism. As the great neurophysiologist Wilder Penfield wrote at the end of his life: "What a thrill it is, then, to discover that the scientist, too, can legitimately believe in the existence of the spirit!"[4]

■ Beyond Naturalism

Given the challenge of naturalism, how can Christians meet this challenge? As I explained in chapter 1, in one sense this challenge is not new; it is the oldest challenge of all. Humanity has always been tempted to trust in nature and make nature an absolute rather than trusting in God. So it is not likely that this challenge will ever disappear, this side of the *parousia*.

The Christian answer to this, ever since the day of the first Pentecost, has been what the New Testament calls *metanoia*. *Metanoia* is loosely translated as "conversion," but a more literal translation is "to change one's mind or outlook." This is what is required to become a Christian. After *metanoia*, what once seemed most trustworthy and stable is no longer, and what once seemed like a fairy tale, God's power and grace, now becomes the most real. Usually this process takes a long time, and there is no magic formula by which it can be achieved.

The great Jesuit theologian Bernard Lonergan spoke of three types of conversion: intellectual conversion, moral conversion, and religious conversion.[5] Intellectual conversion means that knowing is more than just looking and sensing; reality is greater than what is known by the senses; we can know what we cannot see. Moral conversion means the recognition that what is right is other than and greater than one's own wishes. Religious conversion means falling in love with God so that the center of one's life is no longer oneself, but God. We can see these take place in the life of St. Augustine as told in his *Confessions*. First he struggles to understand God as a nonmaterial reality. Once he realizes that God is a spirit, he comes to an understanding of moral conversion—God is the wholly good and beautiful—and then of religious conversion, when the center of his will is shifted from his own natural desires—for wealth, sex, and fame—to love of God.

This book is primarily about intellectual conversion, which is but a part of the total conversion asked of those who would follow God. But it is a necessary prelude to religious conversion. Intellectual challenges have to be met intellectually, and not simply with choruses of religious affirmation. I have tried to show that Christians can explain the evidence of science as well as, and even better than, naturalists. This is at least a start toward conversion.

But of course there is more to conversion than that. Every Christian is expected to live by the power of God's grace and to exhibit that power in his or her life. It is grace that really delivers us from the confines of our natural limitations, our natural fears and inadequacies. Grace gives us the power to overcome our limitations and do things that we thought impossible. Allow me to give a personal example. After college, I wanted to go to graduate school but could not because after four or five hours of reading, I could not concentrate anymore, and this blockage would last for a day or more. So, because of an opportunity, I went into business (commercial waterproofing and roofing) and was quite successful, even though it was not what I wanted to do. I took various courses at night, but still found after ten years that I could not study for the extended periods of time that are required for graduate school. Nothing I did—strenuous exercise, meditation, having a drink—improved this condition. But one Sunday, in church, it came to me that I could go to graduate school. So I took a leave from my business and matriculated at Marquette in the master's program in theology, sixteen years after graduating from college. Classes started on a Wednesday, but by Friday I realized that the same problem was recurring. I was studied out, and thought I would have to leave the program. That Sunday I went to Mass and found, after that Mass, that the blockage had disappeared. I could study all day, all week, without difficulty. I switched to the doctoral program and went through graduate school easily. What I experienced was grace—a gift from God that allowed me to do what I could not do myself and that raised me beyond my natural inadequacies. It was not, though, as if grace suppressed my nature; rather, it elevated and perfected my natural abilities. It is this same power that transforms the lives of alcoholics when they find that by surrendering their problem to a higher power they can be free from alcohol (or other addictions).

It is because I have experienced the power and love of God in my life, and continue to do so, that I find so-called "scientific" explanations of religion unconvincing. I am not a Christian because I expect great rewards in heaven, or because Christianity helps me to bond to a community, or because I am afraid of dying, or because I feel that I have to propitiate an angry God, or for any other extrinsic reason. I am a Christian because I experience in my life the love and presence of God and Jesus Christ now. I pray often because it brings me into God's presence, not because I want to get merit points in heaven.

The greatest gifts of grace are faith, hope, and the love of God (1 Corinthians 13), which, Paul tells us, is poured into our hearts by the Holy Spirit that is given to us (Rom. 5:5). It is this love that allows us to love others, even enemies, and that characterizes the converted Christian life. Such a love is beyond our natural abilities. Naturally, biologically, we tend to love our kin and our friends, those on our side, but not those who are against us. By living according to this love, we give witness to the power of God's grace in our everyday lives and show that we are not bound by the limitations of our genes or even our upbringing.

Even the poorest, most abandoned person can experience the trans-forming power of God's grace. This is the good news of the gospel—not just that we will be happy with God in the afterlife, but that we can be happy with God right now, however desperate our situation. Christianity is not about rules and laws, guilt and fear of punishment, or extrinsic rewards. It is about grace: the experience of God's transforming love and power in our lives that elevates and perfects our natural abilities and allows us to do more than we thought possible. In this sense, the life of every fully converted Christian moves beyond naturalism. It is God's grace that makes the Christian practice of everyday life possible. And it is this same power of grace that one day will bring us to the resurrection, the ultimate transformation of nature, and to eternal life with God.

Notes

Chapter 1: Introduction: The Dying of God

1. Philip Yancey, "God's Funeral" (*Christianity Today* [9 September 2002]: 88).

2. Peter Berger, *The Sacred Canopy* (Garden City: N.Y.: Doubleday, 1967; Anchor Books edition, 1969), 108.

3. See "The Fight for God" (*The Economist* [21 December 2002]: 32–36). About 47 percent of Americans go to a religious service once a week; in Western Europe, about 20 percent; in Eastern Europe, 14 percent.

4. See Agnieszka Tennant, "The Good News About Generations X and Y" (*Christianity Today* [5 August 2002]: 40–45).

5. Berger, *Sacred Canopy*, 129.

6. Bede Griffiths, *Cosmic Revelation* (Springfield, Ill.: Templegate, 1983), 24.

7. Religion includes many facets: teachings, devotion, mysticism, ritual, ethics, law, art, and theology. Theology is the attempt within any religion to understand and explain, through reason, the beliefs and practices which are germane to that religion.

8. Edward O. Wilson, *On Human Nature* (Cambridge, Mass.: Harvard University Press, 1978), 192.

9. There is a slight difference between naturalism and materialism. Some, who claim that only nature exists, still see religious elements in nature. They are often called "religious naturalists" (see note 13 below). Those who call themselves materialists, however, do not associate religious elements with matter; there is no group of "religious materialists" that I am aware of.

10. Bertrand Russell, "A Free Man's Worship," (December 1903), reprinted in *Mysticism and Logic* (U.K.: 1917; Garden City, N.Y.: Doubleday, Anchor paperback, n.d.), 45.

11. In John Bockman, *The Third Culture* (New York: Touchstone Books, 1996), 10.

12. Richard Dawkins, *River out of Eden* (New York: Basic Books, 1995), 133.

13. One can claim that nature is God and God nature; this is the position of pantheism. A somewhat similar position is "religious naturalism," which holds that only nature and natural causes exist, but also holds that there are sacred or religious aspects of nature. See Jerome A. Stone, "Varieties of Religious Naturalism," in *Zygon* 38 (March 2003): 89–93. But such a position still denies the existence of a transcendent God, who is the source and author of nature, and who would exist even if nature did not exist.

14. Jacques Monod, *Chance and Necessity* (New York: Random House, Vintage Books, 1971), 180.

15. See Edward Larson and Larry Witham, "Scientists and Religion in America" (*Scientific American* [September 1999]: 88–93).

16. Kenneth Miller, *Finding Darwin's God* (New York: Harper Collins, 1999), 15, 19.

17. John Searle, *Mind, Language, and Society* (New York: Basic Books, 1998), 35.

18. John Hick, *The Fifth Dimension* (Oxford, U.K.: One World, 1999), 14.

19. Rudolf Bultmann, *Jesus Christ and Mythology* (New York: Charles Scribner's Sons, 1958), 16.

20. Will Provine, "Evolution and the Foundation of Ethics" (*MBL Science* 3 [1988]: 25–29), cited in Miller, *Finding Darwin's God,* 171–72.

21. Marvin Minsky, *A Society of Mind* (New York: Simon and Schuster, 1985), 306–307.

22. See J. L. A. Garcia, "Professor of Death: Peter Singer and the Scandal of 'Bioethics'" (*Books and Culture: A Christian Review* [September–October 2001]), at http://www.christianitytoday.com/bc/2001/005/11.30.html. Also, Peter Singer, *Practical Ethics* (Cambridge, U.K.: Cambridge University Press, 1979).

23. Stephen Jay Gould, *Wonderful Life* (New York: Norton, 1989), 291.

24. Isaac Newton, *Mathematical Principles of Natural Philosophy,* in *Great Books of the Western World* (Chicago: University of Chicago Press, 1952), 34:1, emphasis added.

25. Ibid., 369.

26. Francis Crick, *The Astonishing Hypothesis: The Scientific Search for the Soul* (New York: Charles Scribner's Sons, 1994), 3.

Chapter 2: God and Nature: Historical Background

1. Bernhard Anderson, *Understanding the Old Testament* (Englewood Cliffs, N.J.: Prentice-Hall, 1966), 106. Anderson gives a lengthy account of the struggle between the Yahwist and Canaanite religion in early Israel.

2. John L. McKenzie, S.J., "Miracle," in *Dictionary of the Bible,* ed. John L. McKenzie (New York: Macmillan, 1965), 578.

3. See Raymond Brown, "The Gospel Miracles," in *New Testament Essays* (Milwaukee: Bruce, 1965), 168–91.

4. See Alan Padgett, "The Body in Resurrection: Science and Scripture on the Spiritual Body," in *Word and World* 22/2 (2002), 155–63.

5. Cited in A. O. Lovejoy, *The Great Chain of Being* (Cambridge, Mass.: Harvard University Press, 1936), 63. Lovejoy's work is one of the standard works on hierarchy of being.

6. Origen, *Against Celsus* 4:78, cited in H. Paul Santmire, *The Travail of Nature* (Philadelphia: Fortress, 1985), 50.

7. Cited in Ian Bradley, *The Celtic Way* (London: Darton, Longman, & Todd, 1993), 35. Much of this section on Celtic Christianity is drawn from Bradley's book.

8. Alexander Carmichael, *Carmina Gadelica* (Edinburgh: Oliver and Boyd, 1928–1954), 1: 39–41; cited in Bradley, *Celtic Way,* 58–59.

9. H. J. Massington, *The Tree of Life* (London: Chapman & Hall, 1943), 37; cited in Bradley, 36–37.

10. Bradley, *Celtic Way,* 40, 44.

11. Matthew Fox, *Original Blessing* (Santa Fe, N.M.: Bear & Co., 1983), 316, 76.

12. Augustine, *Confessions* 7.17.23, trans. Henry Chadwick (New York: Oxford, 1992), 127.

13. Augustine, *De Libero Arbitrio* 16.41, cited in Martin D'Arcy, "The Philosophy of St. Augustine," in *St. Augustine: His Life, Age, and Thought* (New York: Meridian, 1957), 166.

14. Augustine, *Sermons* 241.2, cited in Santmire, *Travail,* 66–67.

15. Augustine, *City of God* 22.29, cited in Santmire, *Travail,* 65.

16. Santmire, *Travail,* 70.

17. The Italian word *per* can mean "for," "by," or "through."

18. From *Francis and Clare: The Complete Works,* trans. and intro. Regis Armstrong, O.F.M, and Ignatius Brady, O.F.M. (New York: Paulist Press, 1982), 38–39.

19. M. –D. Chenu, O.P., *Nature, Man, and Society in the Twelfth Century* (Chicago: University of Chicago Press, 1968), 1–48.

20. St. Thomas Aquinas, *Compendium of Theology* 139, in *The Light of Faith: The Compendium of Theology* (Manchester, N.H.: Sophia Press, 1993), 157.

21. Aquinas, *Summa Contra Gentiles* 3.1.71 (Notre Dame, Ind.: University of Notre Dame Press, 1975), 237.

22. See Norris Clarke, S.J., "The Meaning of Participation in St. Thomas," in *Explorations in Metaphysics* (Notre Dame, Ind.: University of Notre Dame Press, 1994), 89–101.

23. See Aquinas, *Summa Contra Gentiles* 4.11, 80–81.

24. See Clarke, "Meaning of Participation," 89–101.

25. Aquinas, *Summa Theologiae* 1.47.a.1.

26. Aquinas, *Light of Faith* 148, 167.

27. See Santmire, *Travail,* 97–119.

28. Zachary Hayes, *The Hidden Center: Spirituality and Speculative Christology in St. Bonaventure* (New York: Paulist Press, 1981), 13.

29. Cited by Leonard Bowman, "The Cosmic Exemplarism of Bonaventura" (*Journal of Religion* 55, 2 [April 1975]: 191).

30. This section is a condensation of a more extensive treatment in Terence Nichols, *That All May Be One: Hierarchy and Participation in the Church* (Collegeville, Minn.: Liturgical Press, 1997), 171–77.

31. From the introduction to the *Expositio Aurea*, cited in Stephen Ozment, *The Age of Reform* (New Haven: Yale University Press, 1980), 58.

32. Erwin Iserloh, in Hans Georg Beck, et al., *From the High Middle Ages to the Eve of the Reformation*, vol. 4 *Handbook of Church History*, ed. Hubert Jedin and John Dolan (Montreal, Quebec: Palm Publishers, 1970), 350.

33. Heiko Oberman, *The Harvest of Medieval Theology* (Cambridge, U.K.: Cambridge University Press, 1963), 255.

34. See Ian Barbour, *Religion and Science* (San Francisco: Harper, 1997), 9–11.

35. Galileo Galilei, "The Assayer," in *Discoveries and Opinions of Galileo*, trans. Stillman Drake (Garden City, N.Y.: Doubleday, 1957), 274.

36. Gary Deason, "Reformation Theology and the Mechanistic Conception of Nature," in David Lindberg and Ronald Numbers, eds., *God and Nature* (Berkeley: University of California Press, 1986), 169.

37. Robert Boyle, *Some Motives and Incentives to the Love of God* (1648), in *Works*, vol. 1, Thomas Birch, ed. (London, 1744), 167, cited in Norma Emerton, "Arguments for the existence of God from nature and science," in Murray Rae, Hilary Regan, and John Stenhouse, eds., *Science and Theology: Questions at the Interface* (Grand Rapids: Eerdmans, 1994), 76.

38. Isaac Newton, *First Letter to Richard Bentley* (1692), in *Newton's Philosophy of Nature*, H. S. Thayer, ed. (New York: Hafner, 1953), 46, cited in Emerton, "Arguments," 78.

39. T. R. Malthus, *An Essay on Population* (London: J. M. Dent, 1960–61).

Chapter 3: God and Nature: Unity or Dissociation?

1. Karl Rahner, *Hominisation* (New York: Herder and Herder, 1965), 53–61.

2. The great Danish physicist Niels Bohr developed a theory of complementarity based on the particle and wave characteristics of subatomic particles. See Niels Bohr, *Atomic Physics and Human Knowledge* (New York: Wiley, 1958).

3. W. Norris Clarke, *The One and the Many* (Notre Dame, Ind.: University of Notre Dame Press, 2001), 44–59. Clarke explains that a true analogy is not like a metaphor. In a true analogy, ". . . the same attribute can be asserted with literal truth of all its analogates . . ." (p. 50). For example, we speak of God as loving, wise, personal, etc., and apply these attributes infinitely to God and in a limited sense to humans. It is literally true of God that he is loving, wise, personal, etc., but it is not literally true of God that he is a rock or a fortress, etc.; these are all metaphors. We may think of God as possessing some rock-like or fortress-like qualities, but it is not true to say that God is a rock or a fortress.

4. See W. Norris Clarke, "Analogy and the Meaningfulness of Language about God," in *Explorations in Metaphysics* (Notre Dame, Ind.: University of Notre Dame Press, 1994), 123–49. The idea that a term such as "being" can be applied to God and to creatures univocally, that is, with exactly the same meaning, is a position often attributed to John Duns Scotus. Clarke distinguishes the Thomistic understanding of the analogy of being from this latter position. Being, he argues, is not only infinitely greater in God, it exists in God in a different mode than it does in creatures. See Clarke, in the article cited above; also, *The One and the Many*, 52. I am following Clarke's understanding of analogy in this book.

5. See W. Norris Clarke, "The Meaning of Participation in St. Thomas," in *Explorations in Metaphysics* (Notre Dame, Ind.: University of Notre Dame Press, 1994), 89–101.

6. Thomas Aquinas, *In Boeth De Hebdom.*, lect. 2, ed. Mandonnet, in *Opuscula Omnia* (Paris, 1927), 1:172–73, cited in Clarke, "Meaning of Participation," 99.

7. Thomas Aquinas, Summa Theologiae, I, q. 8, art. 1, (New York: Benzinger, 1947), vol. I, 34.

8. Thomas Aquinas, *Summa Contra Gentiles* (Notre Dame, Ind.: University of Notre Dame Press, 1957), 4:79–81.

9. John Polkinghorne, *Serious Talk* (Valley Forge, Pa.: Trinity Press International, 1995), 80.

10. Steven Rose, *Lifelines: Biology Beyond Determinism* (Oxford, U.K.: Oxford University Press, 1998), 139.

11. Ibid., 138.

12. Brian Goodwin, *How the Leopard Changed Its Spots* (New York: Charles Scribner's Sons, 1994), 40.

13. Jonathan Weiner, *The Beak of the Finch* (New York: Knopf, 1994).

14. See Arthur Peacocke, *Theology for a Scientific Age*, enlarged edition (Minneapolis: Fortress, 1993), 53–55. Also, Arthur Peacocke, "The Sound of Sheer Silence: How does God Communicate with Humanity?" in Robert John Russell et al., eds., *Neuroscience and the Person: Scientific Perspectives on Divine Action* (Vatican City State and Berkeley: Vatican Observatory Publications and Center for Theology and the Natural Sciences, 1999), 215–47.

15. The relation of information to form is not easy to define. Etymologically, information derives from the Latin, *informare,* to give form, shape, or character to, to be the formative principle of. "Information," as used in English, has both a subjective and an objective sense. Subjectively, it refers to knowledge in the mind; in this sense, persons have information, things don't. But we also might say: "That book or that message contains a lot of information." Information theory can calculate the amount of information in a message (the so-called Shannon-Wiener measure, H). In this sense information is used objectively. (For an extended discussion of the meaning of information, see Paul Young, *The Nature of Information* [New York: Praeger, 1987]). Following the etymological sense of information as giving form to and being the formative principle of something, I will speak of information as primarily referring to knowledge or instructions, but will also speak of "embodied information" as the equivalent of form. Form in this sense is a regular pattern, and it exists in things, like buildings, atoms and molecules, DNA, organic beings like horses and persons, etc. It also exists in time: we speak of the form of a musical composition, schedules, and so on. Information can specify form, as a blueprint specifies the contours of a building, an atomic sequence and pattern specifies a particular molecule, directions specify a route, laws specify behavior, musical notation specifies a composition, a timetable specifies a schedule, and so on. Conversely, we can think of a building as embodying the information in the blueprint, site location, and building specifications, a molecule as embodying the information which specifies it, and so on.

16. What about the creation from nothing, creation *ex nihilo?* When we read Genesis 1 carefully, it looks as if the waters of chaos were already there, and God merely ordered them. Indeed, the Jewish theologian Jon Levenson has argued that God's act of creating the world is still unfinished, and that the persistence of evil and disorder in creation is a sign that the original chaos in still incompletely ordered (Jon Levenson, *Creation and the Persistence of Evil* [San Francisco: Harper and Row, 1988]). But there is another way to interpret the Genesis texts. That is, that the original waters mentioned in Genesis 1 are a poetic rendition of formlessness, that which lacks all form. But what is complete formlessness? It is no thing. For any thing, even an atom, must have form to be an existing thing. So no thing is nothingness. But nothingness is a highly abstract concept, beyond the conceptual range of the early authors of Genesis. For them, complete disorder was best represented by a flood (frequent in the Mesopotamian valley). Thus I would argue that while the idea of nothingness is not explicit in Genesis, the idea of "no-form-ness" is explicit and was as close as these authors could get to the idea of nothingness. Creation from nothing might be said, then, to be implicit in the text. Its explicit formulation was only drawn out much later in Jewish thought (see 2 Maccabees 7:28).

17. See Ray Kurzweiler, *The Age of Intelligent Machines* (Cambridge, Mass.: MIT Press, 1990), 189–98.

18. Ray Kurzweiler, "Reflections on Stephen Wolfram's 'A New Kind of Science.'" Available at http://kurzweilai.net/meme/frame.html?main=/articles/art0464.html.

19. See T. Nichols, "Transubstantiation and Eucharistic Presence" (*Pro Ecclesia* 11, 1 [Winter 2002]: 57–75).

20. This is said by Aquinas: "God is in things as containing them: nevertheless by a certain similitude to corporeal things, it is said that all things are in God; inasmuch as they are contained by him," *Summa Theologiae I,* Quest. 8, art. 1, Reply to Objection 2, 1:34. However, this is also a position typical of panentheism, which asserts that the world is in God and God is in the world, but God is more than the world. Panentheism rejects pantheism on the one hand, and Deism (God is extrinsic to the world) on the other. Panentheism is associated with process theology (see chapter 9), and espoused by philosophers and theologians such

as Charles Hartshorn, Arthur Peacocke, and Philip Clayton. It is so vague, however, that even Aquinas, by some definitions and based on the passage cited above, might be considered a panentheist. The position I am developing in this chapter also might be labeled panentheist. However, the category of panentheism is so imprecise that I would prefer to avoid it altogether. For a good summary and criticism of panentheism, see Gregory Peterson, *Whither Panentheism?* In *Zygon* 36 (3) 395–405.

21. St. Augustine, *Confessions* X, XXVII, Henry Chadwick, trans. (Oxford, U.K.: Oxford University Press, 1992), 201.

Chapter 4: Origins: Creation and Big Bang

1. For a history of the development of the notion of the expanding universe and the Big Bang theory, see Helge Kragh, *Cosmology and Controversy: The Historical Development of Two Theories of the Universe* (Princeton: Princeton University Press, 1996).

2. Ibid., 150.

3. The story is told in Steven Weinberg, *The First Three Minutes* (New York: Harper Collins, Basic Books, 1977, updated edition 1993), 44–76.

4. The following account is drawn form Weinberg, *The First Three Minutes;* Alan Guth, *The Inflationary Universe* (Reading, Mass.: Addison-Wesley, 1997); George Ellis, *Before the Beginning: Cosmology Explained* (London: Boyars, Bowerdean, 1993); and Stephen Hawking, *A Brief History of Time* (New York: Bantam, 1988). For a technical presentation, see Joseph Silk, *The Big Bang*, revised and updated edition (New York: W. H. Freeman, 1989).

5. See Ellis, *Before the Beginning*, 58ff.

6. According to Einstein's famous equation, $E = mc^2$, matter can be transformed into energy and vice versa. This is what happens in an atomic explosion; a small amount of matter is transformed directly into energy. In the earliest stages of the universe, because the temperature was so hot, energy/matter consisted mostly of radiation (energy), with particles (matter) condensing out, colliding, and thereby being transformed back into energy.

7. Weinberg, *The First Three Minutes,* 154.

8. Seth Borenstein, "Universe Slips into High Gear: Einstein's Fudge Factor Debated," *St. Paul Pioneer Press* (6 January 1999), 1A.

9. Guth, *Inflationary Universe*, ch. 1. See also Paul Davies, *Superforce* (New York: Simon and Schuster, 1984), ch. 12.

10. In the inflationary epoch, the universe expands much faster than the speed of light because in addition to the energy/matter, which is expanding at the speed of light, space also expands, like a stretching sheet of rubber, at much faster than the speed of light.

11. Guth, *Inflationary Universe*, chs. 10–12. See also Gregg Easterbrook, "What Came Before Creation?" (*U.S. News and World Report* [20 July 1998]: 44–52).

12. Guth, *Inflationary Universe,* 275.

13. Ibid., 271–76.

14. Peter Atkins, *Creation Revisited* (Harmondsworth, U.K.: Penguin, 1994), 115.

15. Keith Ward, *God, Chance, and Necessity* (Oxford, U.K.: One World, 1996), 35. Ward provides an elegant and convincing rebuttal to Atkins.

16. Hawking, *Brief History,* 136.

17. For a popular description of these, see Davies, *Superforce.*

18. Ibid., 242.

19. John Leslie, *Universes* (London and New York: Routledge, 1989), 33–36. Leslie provides a whole series of these "coincidences" in ch. 2 ("The Evidence of Fine Tuning") of *Universes.*

20. John Leslie, "How to Draw Conclusions from a Fine-tuned Universe," in Robert John Russell, William Stoeger, and George Coyne, eds., *Physics, Philosophy, and Theology* (Vatican City: Vatican Observatory Press, 1988), 298.

21. Fred Hoyle, "The Universe: Past and Present Reflections" (*Engineering and Science* [November 1981]: 8–12), cited in Owen Gingerich, "Is There a Role for Natural Theology Today?" in Murray Rae, Hilary Regan, and John Stenhouse, eds., *Science and Theology: Questions at the Interface* (Grand Rapids: Eerdmans, 1994), 40.

22. Ward, *God, Chance, and Necessity,* 24.

23. Ellis, *Before the Beginning,* 99.

24. John McKenzie, "Aspects of Old Testament Thought," in Raymond Brown, Joseph Fitzmyer, and Roland Murphy, eds., *The New Jerome Biblical Commentary* (Englewood Cliffs, N.J.: Prentice-Hall, 1990), 1293.

25. Ibid., 1293–94.

26. Thomas Aquinas, *Writings on the Sentences of Peter Lombard,* book 2, distinction 1, question 1, article 2, response. English translation taken from

Steven E. Baldner and William E. Carroll, *Aquinas on Creation* (Toronto, Ontario: Pontifical Institute of Medieval Studies, 1997), 74.

27. Ibid., 74–75.

28. Aquinas, *On the Eternity of the World,* in Baldner and Carroll, *Aquinas on Creation,* 114–22.

29. Bertrand Russell, *Why I Am Not a Christian* (New York: Simon and Schuster, 1957), 6–7.

Chapter 5: Evolution: The Journey into God

1. Charles Darwin, *The Origin of Species, The Great Books of the Western World* 49 (Chicago: William Benton, 1952), 33.

2. National Academy of Sciences, *Science and Creationism,* 2d ed. (Washington, D.C.: National Academy Press, 1999), 9.

3. See Murray Gell-Mann, *The Quark and the Jaguar* (New York: W. H. Freeman, 1994), 23–41.

4. William Dembski, "Signs of Intelligence," in William Dembski and James Kushiner, eds., *Signs of Intelligence* (Grand Rapids: Brazos, 2001), 171–92.

5. Richard Dawkins, *The Blind Watchmaker* (New York: Norton, 1986), 17–18.

6. Ibid., 5.

7. Ibid., 127.

8. Richard Dawkins, *The Selfish Gene* (New York: Oxford University Press, 1989), 127.

9. S. J. Gould and N. Eldredge, "Punctuated Equilibria: The Tempo and Mode of Evolution Reconsidered" (*Paleobiology* 3 [1977]: 115–51).

10. Stephen Jay Gould, *Wonderful Life* (New York: Norton, 1989), 302.

11. Stephen Jay Gould, "Self-Help for a Hedgehog Stuck on a Molehill" (*Evolution* 51, 3 [1997]: 1023).

12. Ibid., 1022.

13. Gould, *Wonderful Life,* 25.

14. Stuart Kauffman, *At Home in the Universe* (New York: Oxford University Press, 1995), 25, italics in original.

15. Brian Goodwin, *How the Leopard Changed Its Spots* (New York: Charles Scribner's Sons, 1994); Mae Won Ho and Peter Saunders, eds., *Beyond Neo-Darwinism: An Introduction to the New Evolutionary Paradigm* (London: Academic Press, 1984); Steven Rose, *Lifelines: Biology Beyond Determinism* (New York: Oxford University Press, 1998).

16. David Depew and Bruce Weber, *Darwinism Evolving* (Cambridge, Mass.: MIT Press, 1995), 18.

17. S. J. Gould, "Darwinism and the Expansion of Evolutionary Theory," (*Science* [216]: 383).

18. Rose, *Lifelines,* 304.

19. Ibid., 309.

20. Robert Wesson, *Beyond Natural Selection* (Cambridge, Mass.: MIT Press, 1991), 146.

21. Ibid., 291.

22. Stanley Jaki, *The Relevance of Physics* (Chicago: University of Chicago Press, 1966), 79–86.

23. James C. Maxwell, "Ether," in *Encyclopaedia Britannica* 9th ed. (New York: Samuel L. Hall, 1878–1895), 8:572.

24. See Michael Denton, *Nature's Destiny* (New York: Free Press, 1998), 19–46.

25. Ibid., 334.

26. Ibid., 336.

27. George Cuvier, *Essay on the Theory of the Earth,* trans. R. Kerr (Edinburgh and London: 1813), 94–95, cited in Denton, *Nature's Destiny,* 329.

28. S. A. Kauffman, *The Origins of Order* (Oxford, U.K.: Oxford University Press, 1993), 53–54.

29. Ibid., 338–39.

30. Niles Eldredge, *Timeframes* (New York: Simon and Schuster, 1984), 128.

31. Ibid., 145.

32. Barry Hall, "Evolution on a Petri Dish" (*Evolutionary Biology* 15 [1982]: 85–150).

33. Kenneth Miller, *Finding Darwin's God* (New York: Harper Collins, 1999), 145.

34. D. J. Futuyma, *Evolution* (Sunderland, Mass.: Sinauer Associates, 1986), 478, cited in Miller, 146.

35. Hall, "Evolution," 143.

36. Ibid.

37. Richard Lenski and John Mittler, "The Directed Mutation Controversy and Neo-Darwinism" (*Science,* 259, 5092 [8 January 1993]: 188ff.

38. See Lee Spetner, *Not by Chance* (New York: Judaica Press, 1998), 139–41.

39. William A. Dembski, *Intelligent Design: The Bridge between Science and Theology* (Downers Grove, Ill.: InterVarsity Press, 1999).

40. Michael J. Behe, *Darwin's Black Box: The Biochemical Challenge to Evolution* (New York: Free Press, 1996).

41. Stephen Jay Gould, *The Panda's Thumb* (New York: Norton, 1980), 19–26.

42. Ian Barbour, "Five Models of God and Evolution," in Robert John Russell et al., eds., *Evolutionary and Molecular Biology* (Vatican City: Vatican Observatory Press; Berkeley, Calif.: Center for Theology and the Natural Sciences, 1998), 419–44.

43. Ilya Prigogine, *Order out of Chaos* (New York: Bantam, 1984), 147–48.

44. See James Gleick, *Chaos* (New York: Penguin, 1987), 27.

45. Prigogine, *Order out of Chaos*, 176.

46. Stuart Kauffman, *At Home in the Universe*, 3–69.

47. Jacques Monod, *Chance and Necessity* (New York: Knopf, 1971), 21.

48. Wolfhart Pannenberg, *Systematic Theology* (Grand Rapids: Eerdmans, 1994), 2:109–15.

49. See Ernst Mayr, "Processes of Speciation in Animals," in *Toward a New Philosophy of Biology: Observations of an Evolutionist* (Cambridge: Harvard University Press, 1988), 367–68.

50. See the interesting proposal of Archbishop Joseph Zycinski in "Beyond Necessity and Design: God's Immanence in the Process of Evolution," CTNS Bulletin 22 (winter 2002): 3–10. Zycinski suggests that God can be thought of as a "divine attractor" who "directs all processes toward a transcendent future and attracts them to divine purposes" (p. 10).

51. Thomas Aquinas, *Summa Theologiae* 1.47.1, English translation by the English Dominicans (New York: Benzinger Brothers, 1947).

52. John Polkinghorne, *Serious Talk* (Valley Forge, Pa.: Trinity Press International, 1995), 105.

53. Paul Davies, "Teleology without Teleology: Purpose through Emergent Complexity," in Russell et al., *Evolutionary and Molecular Biology.*

54. Rudolf Brun, "Does God Play Dice? A Response to Niels H. Gregersen, 'The Idea of Creation and the Theory of Autopoietic Process'" (*Zygon* [March 1999]: 95).

55. Polkinghorne, *Serious Talk*, 83.

56. Robert John Russell, "Special Providence and Genetic Mutation," in Russell et al., *Evolutionary and Molecular Biology*, 191–223.

57. Niels H. Gregersen, "The Idea of Creation and Autopoietic Process" (*Zygon* [September 1998]: 333).

58. Polkinghorne, *Serious Talk*, 76–90.

59. Polkinghorne, *Serious Talk*, 104–109.

Chapter 6: Human Nature: Do We Have Souls?

1. See Helmer Ringgren, *Israelite Religion* (Philadelphia: Fortress, 1966), 242. Ringgren argues that ancient Israelites did believe in afterlife, mainly on the basis of texts forbidding necromancy.

2. James Barr, *The Garden of Eden and the Hope for Immortality* (Minneapolis: Fortress, 1993).

3. Oscar Cullmann, *Immortality of the Soul: or, Resurrection of the Dead?: The Witness of the New Testament* (New York: Macmillan, 1958).

4. Krister Stendahl, "Immortality Is Too Much and Too Little," in *The Bible as Document and as Guide* (Philadelphia: Fortress, 1984), 196.

5. Joel Green, "A Reexamination of Human Nature in the Bible," in Warren Brown, Nancey Murphy, and H. Newton Malony, eds., *Whatever Happened to the Soul?* (Minneapolis: Fortress, 1998), 158.

6. Barr, *Garden of Eden*, 36–37.

7. Ibid., 43–45.

8. See E. P. Sanders, *Judaism: Practice and Belief 63 BCE–66 CE* (Philadelphia: Trinity Press International, 1992), 298–303. Sanders notes that Josephus claimed that both Pharisees and Essenes believed in an immortal soul, whereas the Sadducees rejected this belief. Saunders thinks that most Jews believed in personal survival after death, though this belief was somewhat vague.

9. Cullmann, *Immortality*, 30.

10. Ian Barbour, "Neuroscience, Artificial Intelligence, and Human Nature," in Robert John Russell et al., eds., *Neuroscience and the Person* (Vatican City: Vatican Observatory Publications; Berkeley, Calif.: Center for Theology and the Natural Sciences, 1999), 250.

11. Lynn de Silva, *The Problem of Self in Buddhism and Christianity* (London: Macmillan, 1979), 75, cited in Barbour, ibid., 250.

12. I am aware that there is a great deal of controversy over what sayings can be attributed to the historical Jesus and what sayings should be attributed to either the Christian community after his death, or to the Gospel writers themselves. But for the purpose of this chapter, the distinction is not critical. Even if the historical Jesus did not utter a particular saying, it would have been affirmed by the early Christian community. And it is the views of the New Testament communities that we are concerned with here.

13. Barr, *Garden of Eden*, 113.

14. Ibid., 99.

15. Ibid., 113.

16. "Afterlife," in *Encyclopedia Judaica*, 2:336ff., cols. 337–38.

17. St. Justin Martyr, *Dialogue with Trypho*, cited in Simon Tugwell, *Human Immortality and the Redemption of Death* (Springfield, Ill: Templegate, 1990), 80.

18. "Homo igitur, ut homii apparet, anima rationalis est mortali atque terreno utens corpore." Augustine, *De Moribus Ecclesiae* 1.27.52. Cited in Étienne Gilson, *The Christian Philosophy of St. Augustine* (New York: Random House, 1960), 271.

19. Augustine, *Johan. Evangelium* 19.5.15, my translation, cited in ibid., 271.

20. Augustine, *City of God* 22.24.3.

21. Ibid. 21.13, 21.14.

22. Gilson, *Christian Philosophy*, 49.

23. Augustine, *City of God* 13.16.

24. Ibid. 21.12. English translation by Marcus Dods (New York: Random House, Modern Library, 1950), 783.

25. John Calvin, *Institutes of the Christian Religion* 1.15.2. Translated by Henry Beveridge (Grand Rapids: Eerdmans, 1989, 1995), 160.

26. Barr, *Garden of Eden*, 94–116, esp. 105.

27. The degree to which Descartes separates the mind and the body is usually exaggerated. He does say they are separate substances. But note the following passage: "Nature also teaches me by these feelings of pain, hunger, thirst and so on that I am not only residing in my body, as a pilot in his ship, but furthermore, that I am intimately connected with it, and that the mixture is so blended . . . that something like a single whole is produced." *Sixth Meditation*, in *Meditations*, trans. Laurence J. Lafleur (Indianapolis: Bobbs-Merrill, 1960, 1977), 134.

28. R. M. Hutchins, ed., *The Great Books of the Western World* 49 (Chicago: William Benton, 1952), 252–659.

29. Ian Tattersall, *Becoming Human* (New York: Harcourt Brace, 1998), 164.

30. Ibid., 177.

31. Ibid., 10–11.

32. Ibid., 216.

33. Malcolm Jeeves, "Brain, Mind, and Behavior," in Brown, Murphy, and Malony, *Whatever Happened?* 80.

34. Ibid., 78.

35. Michael Arbib, "Towards a Neuroscience of the Person," in Russell et al., *Neuroscience and the Person*, 81.

36. See John Eccles and Karl Popper, *The Self and Its Brain* (New York: Springer International, 1977). Also, Wilder Penfield, *The Mystery of the Mind* (Princeton: Princeton University Press, 1977).

37. Francis Crick, *Life Itself* (London: Macdonald & Co., 1982), 56.

38. Francis Crick, *The Astonishing Hypothesis: The Scientific Search for the Soul* (New York: Charles Scribner's Sons, 1994), 3.

39. Ibid., 265–68.

40. Ibid., 259.

41. Richard Dawkins, *The Selfish Gene* (Oxford, U.K.: Oxford University Press, 1989), v.

42. Edward O. Wilson, *On Human Nature* (Cambridge: Harvard University Press, 1978), 176.

43. Ian Barbour, "Neuroscience, Artificial Intelligence, Human Nature," in Russell et al., *Neuroscience and the Person*, 271.

44. Ibid., 269.

45. Jeeves, "Brain, Mind, and Behavior," 73–98.

46. Ibid., 78.

47. Roger Sperry, "The New Mentalist Paradigm and Ultimate Concern [part 1]" (*Perspectives in Biology and Medicine* 29, 3 [spring 1986]: 413.

48. Crick, *The Astonishing Hypothesis*, 258–59.

49 John Polkinghorne, "So Finely Tuned a Universe," (*Commonweal* [August 1996], 13).

50. Crick, *The Astonishing Hypothesis*, 32.

51. David Chalmers, "The Puzzle of Conscious Experience" (*Scientific American*, special edition, "The Hidden Mind" 12, 1 [August 2002]: 92–93.

52. Ibid., 93.

53. Francis Fukuyama, *Our Post-Human Future: Consequences of the Biotechnology Revolution* (New York: Farrar, Straus, and Giroux, 2002).

Chapter 7: Human Nature: Embodied Self and Transcendent Soul

1. For a study of emergentism across many areas of science, see Harold Morowitz, *The Emergence of Everything: How the World Became Complex* (Oxford: Oxford University Press, 2002). For applications of emergentism to human persons, see Ian Barbour, "Neuroscience, Artificial Intelligence, and Human Nature: Theological and Philosophical Reflections,"

and Philip Clayton, "Neuroscience, the Person, and God: An Emergentist Account," both in *Neuroscience and the Person: Scientific Perspectives on Divine Action,* Robert John Russell et al., eds. (Vatican City: Vatican Observatory Publications, and Berkeley: Center for Theology and the Natural Sciences, 1999), 249–80 and 181–214, respectively. See also the various authors in Warren Brown, Nancey Murphy, and H. Newton Maloney, eds., *Whatever Happened to the Soul?* (Minneapolis: Fortress, 1998), especially the chapters by Warren Brown and Malcolm Jeeves.

2. Consciousness has several meanings. Steven Pinker, in *How the Mind Works* (New York: Norton, 1997), 134–36, distinguishes three senses. (1) self-awareness, (2) access to information—i.e. we can be conscious of something as opposed to unconscious or not-conscious of it; (3) sentience, that is, the capacity to experience suffering, joy, etc. When speaking of the consciousness of God, all three definitions would apply.

3. "Brain, Mind, and Behavior," in Brown, Murphy, and Maloney, *Whatever Happened?* 87.

4. See Clayton, "Neuroscience, the Person, and God," cited above.

5. Morowitz, *The Emergence of Everything,* 193–94.

6. Nancey Murphy, ". . . downward mental causation also results in the reshaping of the agents neural pathways" in "Neuroscience and Human Nature," Ted Peters et al., *God, Life and Cosmos: Christian and Islamic Perspectives* (Hampshire, U.K.: Ashgate, 2003), 384.

7. This problem was suggested to me by Gregory Peterson, in a personal communication. It is mentioned in his article, "Whither Panentheism," in *Zygon* 36 (3): 395–405.

8. (New York: Bantam, 1975)

9. See the authors listed below in notes 10 and 11, as well as the following: Melvin Morse, M.D., *Closer to the Light* (New York: Ballantine, 1990); Dr. Peter Fenwick, *The Truth in the Light* (New York: Berkeley Books, 1995); Karlis Osis, PhD. and Erlendur Haraldsson, Ph.D., *At the Hour of Death* (Norwalk, Ct.: Hastings House, 1995); Lee Bailey and Jenny Yates, eds., *The Near Death Experience, A Reader* (New York: Routledge, 1996). For a skeptical account, see Susan Blackmore, *Dying to Live* (Buffalo, N.Y.: Prometheus, 1993). Gary Habermas and J. P. Moreland respond to Blackmore's arguments in *Beyond Death* (Wheaton, Ill.: Crossway, 1998), 199–218.

10. Elisabeth Kübler-Ross, *On Children and Death* (New York: Macmillan/Collier, 1983), 207–11.

11. For responses to naturalistic explanations of near-death experiences, see Michael Sabom, *Recollections of Death: A Medical Investigation* (New York: Harper and Row, 1982), 151–78; Raymond Moody, *Life after Life* (New York: Bantam, 1975), 155–77; Gary Habermas and J. P. Moreland, *Beyond Death,* 155–218.

12. Susan Blackmore, "Near-Death Experiences: In or out of the Body?" in Bailey and Yates, eds., *The Near-Death Experience,* 285–97. See also her *Dying to Live: Near-Death Experiences.*

13. See Habermas and Moreland, *Beyond Death,* 186 and references there; also Sabom, *Recollections,* 178.

14. Blackmore, "Near-Death Experiences," 294.

15. Eugene d'Aquili and Andrew Newberg, *The Mystical Mind* (Minneapolis: Fortress, 1999), 121–43.

16. Sabom, *Recollections,* 6–7.

17. Ibid., 84–86.

18. Ibid., 115.

19. Ibid., 184.

20. Malcolm Jeeves, "Brain, Mind, and Behavior," in Brown, Murphy, and Malony, *Whatever Happened?* 87.

21. T. A. Geissman, *Principles of Organic Chemistry* (San Francisco: W. H. Freeman, 1968), 46.

22. Morowitz, *The Emergence of Everything,* 55.

23. Ibid., 56.

24. Ibid.

25. Ibid., 57.

26. See Malcolm Browne, "Far Apart, Two Particles Respond Faster than Light," *New York Times* (22 July 1997), C1.

27. John Polkinghorne, *The Quantum World* (Princeton: Princeton University Press, 1984), 76.

28. John Hitchcock, *Atoms, Snowflakes, and God* (Wheaton, Ill: Theosophical Publishing House, 1986), 144.

29. Richard Lenski and John Mittler, "The Directed Mutation Controversy and Neo-Darwinism" (*Science* 259, 5092 [8 January 1993]: 188).

30. Wilder Penfield, *The Mystery of the Mind* (Princeton: Princeton University Press, 1975), 52.

31. Ibid., 76.

32. Ibid., 76–77.

33. Sharon Begley, "Even Thoughts Can Turn Genes 'On' and 'Off,'" *Wall Street Journal* (21 June 2002), B4.

34. See Karl Rahner, S.J., "Concerning the Relationship between Nature and Grace," in *Theological Investigations*, Vol. I (Baltimore: Helicon Press, 1961), 297-317. See also Karl Rahner, *Foundations of Christian Faith* (New York: Crossroad, 1982), 126-133. Rahner uses the term "existential" to indicate that this supernatural elevation is not intrinsic to pure human nature, but is a super-added gift of God. But Rahner did not think there were in existence any "purely natural" human beings, that is, humans who did not have God as their end. So God's creation of human beings, and God's elevation of human beings to a supernatural end, would seem to be both involved in the emergence of humanity. I would agree, and think of the "creation of the soul" as meaning the coming into existence of a natural, human, embodied soul, with a limited natural horizon, and its elevation into a relationship with God, as parts of a single process. I realize this raises many theological and scientific problems, but to explore all these problems would require another book.

35. Simon Tugwell, *Human Immortality and the Redemption of Death* (Springfield, Ill.: Templegate Publishers, 1990), 76-83.

36. Ian Tattersall, *Becoming Human* (San Diego: Harcourt Brace, 1998).

37. John Polkinghorne, *Serious Talk* (Valley Forge, Pa.: Trinity Press International, 1995), 106.

38. See *Catechism of the Catholic Church* (Mahweh, N.J.: Paulist Press, 1994), # 366, and references there. John Paul II repeated the teaching of Pius XII (in *Humanii Generis*) in his "Message to the Pontifical Academy of Sciences: On Evolution", # 5, 22 October, 1996 (available at: http:www.ewtn.com.library/PAPALDOC/JP961022.HTM).

39. Norris Clarke, *Explorations in Metaphysics* (Notre Dame, Ind.: Univ. of Notre Dame Press, 1994), 102-122. See also John Zizioulas, *Being as Communion* (Crestwood, N.J.: St. Vladimir's Seminary Press, 1993).

40. Karl Rahner, *Hominisation* (New York: Herder & Herder, 1965), 60–61.

41. We can get an approximate idea of how God could be infinitely distant yet infinitely close to humanity by using a number system as an analogy. Let us say that God is infinity, and any creature is represented by a number-say the number of bits of information which would specify its form. However large the number which represents the creature, God will still be infinitely distant from it. But that number is also contained in God, who, as infinity, contains all numbers. Therefore both God and the number (any number) have a kind of unity; they are not wholly different. However, this analogy is limited, and needs to be complemented by other analogies.

42. Partly this depends on what we mean by "wholly other." God is wholly other if we mean by that that God is God's own cause, whereas every creature is caused, i.e. brought into being and held in being by God. In this sense only God is Creator; all other beings, including angels, are created. But if there is no unity at all between God and creatures, then no term we apply to God has any meaning at all—all terms used of God are literally vacuous. And this is emphatically not the Christian tradition. My position is that terms used about God are used analogically, not univocally, and this includes the terms "being" or "exist." On the difference between the analogy of being, and the position attributed to Duns Scotus, that being is used univocally of God and creatures, see Norris Clarke, *The One and the Many* (Notre Dame, Ind.: University of Notre Dame Press, 2001), p. 52.

43. E. O. Wilson, *Sociobiology* (Cambridge, Mass.: Harvard University Press, 1980), 3, cited in Keith Ward, *Defending the Soul* (Oxford, U.K.: One World, 1992), 66.

44. Wilson, *Sociobiology*, 120.

Chapter 8: Miracles: Signs of God's Order

1. Cited in R. Z. Lauer, "Deism," in the *New Catholic Encyclopedia* (Washington, D.C.: Catholic University of America Press, 1967), 4:723.

2. David Hume, *An Enquiry Concerning Human Understanding* (1748), in E. A. Burtt, *The English Philosophers from Bacon to Mill* (New York: Random House, 1939), 656.

3. Paul Davies, "Physics and the Mind of God," *First Things* (August–September 1995), 34.

4. A. Peacocke, "The Sound of Sheer Silence," in Robert John Russell et al., *Neuroscience and the Person* (Vatican City: Vatican Observatory Publications; Berkeley, Calif.: Center for Theology and the Natural Sciences, 1999), 234.

5. *Nature* 310 (1984): 171.

6. Rudolf Bultmann, *Jesus Christ and Mythology* (New York: Charles Scribner's Sons, 1958), 16.

7. Malcolm Browne, "Far Apart, Two Particles Respond Faster than Light," *New York Times* (22 July 1997), C1.

8. Donald MacKay, Letter to the Editor (*Nature* 311 [1984]: 502).

9. The type of God's action in a miracle is what theologians call special providence, as distinct from God's general providence, by which God guides all events universally, and God's *creatio continua*, the continuous sustaining of every created thing.

10. Alexis Carrell, *The Voyage to Lourdes* (New York: Harper & Brothers, 1950), 22.

11. P. G. Boissarie, *Healing at Lourdes* (Baltimore: John Murphy Co.), 197.

12. Ibid., 199.

13. Ibid., 199.

14. Ibid., 2–9.

15. See, for example, François Leuret and Henri Bon, *Modern Miraculous Cures* (New York: Farrar, Straus, and Cudahy, 1957).

16. Quoted in Louis Monden, *Signs and Wonders* (New York: Desclee, 1966), 195.

17. Alexis Carrell, *Man the Unknown* (New York: Harper and Brothers, 1935), 149.

18. Monden, *Signs and Wonders,* 235.

19. R. M. Grant, *Miracle and Natural Law in Graeco-Roman and Early Christian Thought* (Amsterdam, Netherlands: North Holland, 1952), 214.

20. John Polkinghorne, *The Faith of a Physicist* (Princeton: Princeton University Press, 1994).

21. John Polkinghorne, *Serious Talk* (Valley Forge, Pa.: Trinity Press International, 1995), 76–90.

22. Brian Grene, *The Elegant Universe* (New York: Random House/Vintage Books, 1999).

23. Paul Davies, "What Are the Laws of Nature?" in John Brockman, ed., *Doing Science: The Reality Club* (New York: Prentice-Hall, 1991), 60.

24. St. Augustine, *Reply to Faustus the Manichaean,* in Philip Schaff, ed., *The Nicene and Post-Nicene Fathers,* first series (Peabody, Mass.: Hendrickson, 1995), 4:321–22.

25. Wolfhart Pannenberg, *Systematic Theology* (Grand Rapids: Eerdmans, 1994), 2:83.

26. Robert John Russell, "Special Revelation and Genetic Mutations: A New Defense of Theistic Evolution," in R. J. Russell, William Stoeger, and Francisco Ayala, eds., *Evolutionary and Molecular Biology* (Vatican City: Vatican Observatory Publications; Berkeley: Center for Theology and the Natural Sciences, 1998), 191–223.

27. See Brian Grene, *The Elegant Universe.*

28. (Albany: SUNY Press, 2000), 11.

29. See Nima Arkani-Hamed, Savas Dimopoulos, and George Dvali, "The Universe's Unseen Dimensions" (*Scientific American* [August 2000]: 62–69).

30. Polkinghorne, *Serious Talk,* 108.

Chapter 9: Christianity and Science: Conflict or Complementarity?

1. *Science and Creationism,* 2d ed. (Washington, D.C.: National Academy Press, 1999), 1.

2. George Ellis, *Before the Beginning: Cosmology Explained* (London, New York: Boyars/Bowerdean, 1993), 15–16.

3. Ibid., 17.

4. Arthur Eddington, *The Nature of the Physical World* (Cambridge: Cambridge University Press, 1928), 16.

5. Michael Polanyi, *Personal Knowledge* (Chicago: University of Chicago Press, 1958).

6. Kenneth Miller, *Finding Darwin's God* (New York: Harper Collins, 1999), 208.

7. St. Anselm, *Proslogium or Faith Seeking Understanding,* in *Basic Writings,* S. N. Dane, trans. (LaSalle, Ill.: Open Court, 1962, 2d. ed., 1979), 1.

8. John Polkinghorne, *Science and Theology: An Introduction* (Minneapolis: Fortress, 1998), 98.

9. Walter M. Abbott, S.J., *The Documents of Vatican II* (n.c.: American Press, 1966), 116.

10. John Polkinghorne, in an unpublished address to the University of St. Thomas, April 9, 1992.

11. See Polkinghorne, *Science and Theology.*

12. See John Haught, *Science and Religion: From Conflict to Conversation* (New York: Paulist Press, 1995).

13. Ian Barbour, *Religion and Science* (San Francisco: Harper, 1997), 77–105.

14. Stephen Jay Gould, *Rocks of Ages* (New York: Random House/Ballantine, 1999), 6.

15. John Polkinghorne, *Serious Talk* (Valley Forge, Pa.: Trinity Press International, 1995), 7.

16. *Alcoholics Anonymous* (New York: Alcoholics Anonymous World Services, Inc., 1976 and later editions).

17. David Ray Griffin, *Religion and Scientific Naturalism: Overcoming the Conflicts* (Albany: SUNY Press, 2000), 11, 15.

18. A. N. Whitehead, *Process and Reality* (New York: Macmillan, 1929), 526.

19. Ibid., 522.

20. Ibid., 523.

21. W. Norris Clarke, *Person and Being* (Milwaukee: Marquette University Press, 1993), 20–21.

22. One of the most successful syntheses of process thought and Christian theology is by Joseph Bracken S.J., *The One in the Many* (Grand Rapids: Eerdmans, 2001). Bracken, however, modifies some features of Whitehead's thought. See also C. Robert Mesle and John Cobb Jr., *Process Theology: A Basic Introduction* (St. Louis: Chalice Press, 1993). Other authors influenced by process thought are Ian Barbour and John Haught (see his *God After Darwin: A Theology of Evolution* [Boulder, Co.: Westview Press, 2000]).

23. John Polkinghorne, *Belief in God in an Age of Science* (New Haven: Yale University Press, 1998), 56.

24. See L. R. Lucas and P. E. Hodgson, *Spacetime and Electromagnetism* (Oxford: Oxford University Press, 1990).

25. A. N. Whitehead, *Science and the Modern World* (New York: Macmillan/Mentor Book, 1960), 19.

26. Ibid., 19.

27. Stanley Jaki, *The Road of Science and the Ways to God* (Chicago: University of Chicago Press, 1978).

28. Ian Barbour, *Religion and Science* (San Francisco: Harper, 1997), 24–29, 90.

Chapter 10: Conclusion: Beyond Naturalism

1. John Hick, *The Fifth Dimension* (Oxford, U.K.: One World, 1999), 14.

2. Ian Tattersall, *Becoming Human* (San Diego: Harcourt Brace, 1998), 10.

3. George Ellis, *Before the Beginning: Cosmology Explained* (London: Boyars/Bowerdean, 1993); John Polkinghorne, *Serious Talk* (Valley Forge, Pa.: Trinity Press International, 1995); Polkinghorne, *The Faith of a Physicist* (Princeton: Princeton University Press, 1994).

4. Wilder Penfield, *The Mystery of the Mind* (Princeton: Princeton University Press, 1975), 85.

5. Bernard Lonergan, *Method in Theology* (New York: Seabury, 1972), 238–43.